# READINGS IN WOOD

# READINGS IN WOOD

## What the Forest Taught Me

*John Leland*

*The University of South Carolina Press*

© 2015 University of South Carolina

Published by the University of South Carolina Press
Columbia, South Carolina 29208

www.sc.edu/uscpress

Manufactured in the United States of America

24 23 22 21 20 19 18 17 16 15     10 9 8 7 6 5 4 3 2 1

Library of Congress Cataloging-in-Publication Data
can be found at http://catalog.loc.gov/

This book was printed on recycled paper with
30 percent postconsumer waste content.

ISBN 978-1-61117-458-8 (pbk)
ISBN 978-1-61117-459-5 (ebook)

For Isa, Edward, and their children

# Contents

# CONTENTS

# INTRODUCTION

When but a child, I learned that our ancestors came out of the trees, stood upright on the savannahs, and became human. Which means there is much in me—and perhaps you—that is something less than human. Because powerful desires urge me to return to the trees. Climbing some great-trunked monarch, who has not thought herself the first to scale such heights? Unsteady on a wind-shook branch, not looked like Balboa with wild surmise upon a world beyond the world she knew, be it but the view over the backyard fence? Grown and inhaling the sweet scent of a pine or cedar Christmas tree, not flown through space and time to a better time and space? Not fallen amorous and hugged a shaggy lichened bark? Not kissed a wounded trunk and tasted tree blood on her lips? Not sat long sunk in thought watching a forest burn within her fireplace?

Greater minds than mine have sought in trees answers to profounder questions than I have thought to ask the blue-bleak embers of a dead tree's dying fire. Do you, they ask, like us, rejoice in sunny days, dance with the wind, and blush to have your sexual desires known by prurient passersby? Why, like us, do you torture yourselves reaching for a heaven beyond your grasp? Why twist round yourselves so that your grain becomes a record of your grief? What mystic patterns of science, math, religion hide in your whirls of leaf and branch?

As always, my inspiration is my son, Edward, whose questions are unfailingly impossible to answer. Fortunately, there are grown people wise enough to ask many of the same questions, and I have taken to the trees the guides that others wrote and learned, leaf by leaf, the wisdom of wood. Tom Wessels taught me much of what we can read of a forest's past, Annie Dillard of nature's spiritual side, Wendell Berry of the power of place. Countless on-line botanists explained the particulars of this plant or fungus, geologists this rock, historians this place, mystics this meaning. And the people of the Shenandoah with whom I've walked and talked have taught me more than all of these, the depth that roots, human and vegetable, can sink when given time. And slowly, slower than the slowest student, I have tried to read, mark, and learn lessons both practical and spiritual. I'm still teacher enough to succumb to the temptation to lecture on a stump. Such sawing can but dull the sharpest mind. And what's the point of practicality unless it point the way to something more profound? I don't count tree rings to practice numbers; I count to connect with something other than myself.

In wildness, Thoreau thought, salvation lies. And it's my soul's salvation I seek among the trees, whose gospels botanical Luthers have translated for this ill-educated man who reads but human tongues. Jesus took Himself for forty days into the Wilderness to sort things out; surely I can spare an hour or two a week to try to translate the cryptic Chaldee of the forest's record of birth and death and resurrection. *Readings in Wood* comprise dictation taken from the spirit moving over the face of the deep, dark forest that rings the valley in which I find myself, a mongrel testament of science, faith, superstition, and disbelief I have learned from sitting on tree trunks. *Book* descends to us from Old English *boc,* cousin to *bece,* beech tree. Linguists believe the Angles and Saxons living in the woods of northern Europe made *boc* from *bece,* cutting the magic letters of their runic alphabet in the wood of the same beech tree lovers carve initials in today. These leaves you desultorily turn over once hung in a green wood gone to make this book. Touching a book, you touch a tree. I pray that *Readings in Wood*'s essays, touching you, may justify in some small way the trees that died in their making.

# AMONG THE GRAVES OF TREES

The shade here is thick, the trees so many and large trunked, you can, if willing, pretend you do not know. But I know, and I will make you know: we are walking a graveyard. Eighty years or so ago, strong men came with their crosscut saws and felled a forest older than the oldest white man's roots in these parts. If so inclined, you can drive hours to find scraps and remnants of the Great Eastern Forest, Longfellow's "forest primeval" of "murmuring pines and hemlocks," Gordon Lightfoot's "deep, dark forest . . . too silent to be real." But real it was. And you can trace its closer trace in the wide-planked floors and thick-beamed ceilings of Valley cabins and the weathered post-and-beam cathedrals of bank barns.

And here in the rich damp hollows of the mountains you can still see its last stumps not yet sunk in leaf mold and decay. When every fourth tree was a chestnut, one hundred feet tall, ten to fourteen feet in diameter, a cornucopia fed and housed a nation. In a once upon a finer time than ours, chestnuts roasted annually on open fires, chestnut leaves shaded the village smithy, chestnut wood made up the beams, floors, walls, and furniture of half the country's houses, chestnut ties supported thousands of miles of rail tracks, and chestnut poles held aloft the telegraph lines carrying the SOS that spelt the trees' end, a blight that killed four billion trees in four decades. Accidentally introduced to the New World sometime around 1900 on nursery trees imported from Asia, the blight's evil agent is a fungus (*Cryphonectria parasitica*) that attacks both American chestnuts and their close kin, the chinquapin. Invading through wounds, the fungus wraps its way beneath the bark, around the trunk, its canker strangling the tree, killing its cells with oxalic acid, the same compound that makes wood sorrel such a delight to nibble.

Ghost forests raised their blanched white limbs toward deaf heaven as a younger, poorer forest replaced them. And then the ghosts fell one by one, unheard by a forgetting world, until all that the blight left were weathered barns and memories and the carcasses of behemoths larger and more numerous than we can believe. My father showed me these forests of the dead when we took a bus trip through the mountains. Young and indifferent though I then was, I still remember the profound sadness of a mountainside littered with the thrown trunks of more trees than I could count.

Like many hardwoods, the chestnut has an outer shell that resists rot better than its heart. Long lying on the slopes, these giants rotted from within, consumed themselves, and left like sea shells their hard exteriors as a memory. Not that all chestnuts ended on the forest floor. When the end came, those strong men with saws felled the ghostly forests in an attempt to salvage something from disaster. Blighted, bored and burrowed by bugs, the lumber sold as wormy chestnut. Just last week, I sat in a study paneled with such chestnut, whose occupant informed me, a little too proudly perhaps, that there is no getting more of it.

Yet humans are not alone in having hope of resurrection, and the forest understory hides thickets of chestnut saplings sprouting from the still-living roots of the long-gone giants of the past. They sprout, they flourish, and then the blight, whose curse is everywhere, in the dirt we tread, the air we breathe, returns and kills, garroting the cambium. Next year still there comes forth a shoot from the stump, and a branch grows out of his roots. And again the invisible and pathogenic slay it. And on and on the battle goes, until a coppice of the dead surrounds the buried root.

So enduring is chestnut that you can still find an odd stump here and there, hollow, gray, and brittle with age. Kneel and touch with reverence the relic of a forest gone forever.

All hollowed stumps need not be hallowed, though. You may in ignorance begot of faith be paying devotion to an oak. Rotting inwardly, many a noble hardwood standing now is but a hollow shell of decay. A broken branch, a snag—and water, fungus, beetles, grubs, whatever spreads rot creeps slowly groundward. Knock on the trunk; does it ring sounder than the laminated coins of our day? Or hollow as a false friend? As children, we sought such trees to drum upon and make reverberate with stick and stone and danced around them to the beat of our own drums. Now though I walk with a cane through forests barely older than myself, I still find time to be a perambulating woodpecker and rap upon a likely hollow instrument my song to a world that doesn't listen.

Lacie collects her potting soil from such whited sepulchres, scooping from their trunks their own dead flesh, a rich red brown stew of loam that only yesterday was solid wood, was living plant when she were wore diapers. It is just as well that we cannot speak vegetable, because they suffer torments that understanding would drive us mad. I have seen an acorn taken root in the rot spilled from the tree that gave it birth, a silent Madonna feeding with her flesh her child. On Thunder Ridge a wild cherry seed rooted in the fallen trunk of another of its kind, on whose quick rotting wood it fed, the new tree's roots hardening round and down the fallen trunk until the cherry now handstands, its roots splayed out a foot above the grave of what foresters term in cruel irony a nursery tree. Whose long-gone diameter the cherry's vacant arc remembers—twelve inches. Nor need we wander woods to see the cannibal. We had an ancient pear tree in whose crotch a cedar grew. A hackberry by the look of it has built a tree house in a basswood beside Washington and Lee University's colonnade. In Florida I've seen strangler figs enveloping the tree that nurseried them in roots that turn to trunks. Though less spectacularly, here too trees strangle. Asian honeysuckle, rampant and untamable, embraces soft-wooded trees so tightly she carves a winding channel up their trunks lasting years after the vine vanishes. Always to the right, a dextral spiral staircase, as Charles Darwin observed in 1875. I need to find me a sound one and make a cane of it to rap trees as I hike.

Today's blights are potentially as lethal as the chestnut's. Newer ghost forests haunt the higher slopes of our mountains. The ravenous European gypsy moth (*Lymantria dispar*), whose innumerable caterpillars once defecated so much the falling frass sounded like rain and whose fluttering males were thick as midges, first came to North America as an ill-starred experiment in raising caterpillars for silk. Some escaped, and their descendants, as numerous as the stars in the sky, possessed our Eastern forests where all who know them curse them. There are those who diaper their trees with bands of sticky cloth to catch the wandering caterpillars. But who could diaper a hundred million oak trees from Massachusetts to Carolina? Defoliated, grove after grove died, the leafless, twigless branches clawing a deaf heaven. Already their successors have grown a green bank round their bases. When old, our children may walk their shade, but you and I, should we go mountainward, are damned to stroll between the corpses of those our grandparents' greed for silk destroyed.

The gypsy's golden horde, sated, thinned. And the forest held its breath, awaiting man's next horror: Dutch elm disease. Hold not the Dutch responsible; their scientists merely identified the fungus responsible (one of three *Ophiostoma* species). Brought in on infected logs, the Eurasian fungus is spread by native beetles, whose larvae tunnel beneath the elm's bark. Attempting to halt the fungal spread, the infected elm plugs its own xylem tubes, preventing as well the transport of nutrients, and kills itself. To halt the spread of fungus-bearing beetles, arborists used to spray the popular street trees with a toxic stew of DDT and other poisons, saving some trees but indiscriminately killing any and all insects—and the birds that fed on them, as Rachel Carson warned in *Silent Spring*. A chastened North America and Europe ceased spraying and listened as trees fell, in city and forest. Today, you must drive to Canada, to Winnipeg, Manitoba, to see elms as they were in our grandparents' day. And nowhere will you find forests such as those Herman Melville asks in *Malvern Hill* if they remember the Civil War: "Does the elm wood / Recall the haggard beards of blood? . . . Does Malvern Wood / Bethink itself, and muse and brood?" It is we, instead, haunted by the ghosts of forests gone, who muse and brood.

And kill, and keep on killing. The hemlock wooly adelgid arrived by accident in the 1920s. Named for the innumerable cottony-looking sacs, each with hundreds of eggs, that it attaches to the branches of infected trees, the wooly adelgid (*Adelges tsugae*), small as Lincoln's nose on the penny, sucks the phloem sap of tender new growth, desiccating the needles, which die and fall. As will the tree. As have so many stream-shading hemlocks that the native trout are threatened too with extinction. Where once the hemlocks' deep, dark shade kept waters cold and chilled our souls, harsh sun illuminates streams so choked with debris one wonders how a fish can swim. One by one, the streams I knew as shaded havens in August heat are sun-beaten barrens I ford by clambering from rotting trunk to rotting trunk. Beloved by someone, individual hemlocks can sometimes be saved by a bath of insecticide. But who can wash clean the eastern half of North America? And what would become of our streams and forests were we to rinse them in a Noah's Flood of chemicals? Go out and hike whatever heaven you know, for the hemlocks' Götterdämmerung is nigh, and their end is soon.

And still destruction comes, the tocsin rung now for the ash tree by the emerald ash borer (*Agrilus planipennis*). A recent arrival, this metallic green beetle from northeastern Asia kills when its larvae, feeding invisibly on trees' sapwood, excavate secret galleries beneath the bark, strangling with slow death millions of America's most valuable trees. As with the other plagues, individuals can be spared by injecting powerful toxins tree by tree. The ash in your backyard, perhaps, or those shading the streets you walk. But who will inoculate a forest? Already foresters have hung three-sided sticky traps along the Shenandoah's roads to trap and count the coming infestation, their purple the shade of martyrdom. Go, take your children this summer to a baseball game. For, when the ashes go, the tree that gave Babe Ruth his bat will vanish and another bit of America's soul die.

The gray ghosts of fallen chestnuts I saw as a boy haunt me still. Streets where the elm tree's vase shape rose stand vacant. The Blue Ridge's gnarled and wizened dying oaks stalk my son's dreams. Our backyard hemlocks whiten with adelgids. I have bought for memory's sake an ashen bat. And who knows what other ghosts we have already cursed our children's children with lurk in our forests?

# BY INDIRECTIONS
# FIND DIRECTIONS OUT

The trail took me up through the rocky rubble left from a million and more years of wearing down by a creek reduced to gurgling underground in our drier, warmer days. Bedrock ribbed hillsides gaunt with anorexia. And trees inured to drought—hickories and oaks—had run out the tulips and basswoods growing spoiled in the damp and deep-soiled hollows below. They'd logged here years ago; indeed, the trail was an old logging road built to take the forest out. But here and there, for reasons now unknown, they'd skipped a tree, and these—oaks mostly—stood patriarchal, Abrahams receiving obeisance from a lesser, younger lot.

As weather-beaten and craggy a crew of shepherds you'd meet this side of Genesis, crowns raised in arthritic protest to the gods who strike them for raising limbs toward heaven. Titan after titan lay overthrown, crowns to the northeast, roots to the southwest—from which direction the storm that had destroyed them had come. Long ago enough for those forest denizens who profit from such death to have stripped them of their bark robes and left their naked trunks to bleach in sun and rain. Some, as broken limbed and naked as their comrades, still stood, condemned to bearing witness to the slaughter. And here and there, a younger oak sang requiem when the winds moved in the limbs he'd raised to keep a dead progenitor from falling, unmourned, to the ground.

Trunk after naked trunk betrayed the survival of the fittest that our textbooks describe in terms so bland they bore us. Used as I am to lumber cut from straight-trunked trees we raise like vegetables, these broken ones seemed as torn and tortured as the crucified once hung from trees like these. Bleached skeletons and broken branches spoke of a brief battle between wind and tree. But a war of decades had carved itself into trunk after trunk, whose grains spiraled as if tree after tree had twisted round and round itself, fighting demons none could see.

To what end this tortured turning of xylem and phloem? No one knows, though many speculate. Those who would have invisible forces shape our destinies and bodies believe such corkscrewing the outward and visible sign of the invisible workings of the Coriolis force. Named after the nineteenth-century French engineer and mathematician Gaspard-Gustave Coriolis, the effect describes the tendency of objects on a rotating surface such as a water-wheel or Earth to deflect clockwise with respect to the direction they travel in. So that winds and currents bend to the right in the Northern Hemisphere, to the left in the Southern. Alfred Einstein himself argued that the Coriolis force causes rivers to meander rightwards above the Equator, leftwards below. Which leads me to dream of rivers running in circles, Ouroboroses devouring themselves. That they don't suggests that more localized effects—topography and geology—have a greater effect. Which also explains why, despite urban legend, pig's tails and flushing toilets don't all turn right in Iowa and left in Queensland. But larger hurricanes do swirl counterclockwise off the Outer Banks, and typhoons churn clockwise off Australia's Barrier Reef.

Look closely, and you can sometimes discern the torture of a living tree. Near the Appalachian Trail's shelter on Thunder Hill an oak leans southward, its one-sided crown misshapen and top heavy. On its aching back has grown a compensating ridge as if the bark had split and the long-suffering trunk raised as best it could a callous to protect its cambium and counterweight the pull of gravity. Alongside this ridge, your eyes can see, your fingers trace the bark twisting and spiraling round and under gravity's bent burden. Some claim to have found in Norway's forests that the crowns of trees north of the Equator grow larger on their sunnier, southern sides, that these lopsided weights, coupled with the force of Coriolis, twist tree after tree to the right. Which does not, however, explain my oak's left-turning torture.

Such left-leaning dervishes bedevil those who would patterns find in na-ture. Sinistral trees turn leftward, dextral rightward. My forests of the fallen are, like humans, mostly dextral. As are most of the trees foresters have both-ered measuring. And many bother, because they'd like to learn why some trees twist—and put an end to it so they might raise but the straight and narrow grained. As with so many defects, some blame nature, some nurture, some a combination of the two. That pines are thought to change from left to right as they age perplexes all camps. Nature's partisans offer as proof that certain populations—groves you and I would call them—appear to prefer turning right, others left, and that the angle of turn varies from species to species. Genetics, they declare. But the why or wherefore of tree handedness puzzles foresters as much as does human handness our doctors. Since 70 to 90 percent of us are right-handed, those inquiring wish to know what accounts for such aberrants as Herbert Hoover, Fidel Castro, Pat Robertson, and your humble author. Why don't we go extinct? Scrambled brain hemispheres? A twin that vanished in the womb? The insidious force of Coriolis? There are no randy mailmen among the trees to explain their twisted longings to the wrong.

More nurturing scientists suggest the twisting is a response to stress— prevailing winds, sunlight, an imbalanced crown, or mismatched root and crown. Suppose you are a tree sprung from a rock, your roots groping to the north for sustenance, your crown growing west toward the sun. Some say the cells providing transport must necessarily spiral to connect root and leaf. That prevailing winds may twist a tree as surely as they bonsai the Table Mountain pines that cling precariously to the rocks atop the Shenandoah's ridges. But these, alas, are too thick barked, gnarled, and small to test. And who would cut down a living monument to orneriness to verify a theory? Yet others of an engineering bent suppose such bent grains better able to withstand the torsion of erratic winds. Just as we strengthen ropes by twisting their component strings, so trees might stronger grow if spiraling. Such straightforward-sounding proofs, however, can turn wrongly mathematical: "Consider an infinite cylinder of the radius $R$, fulfilled by an orthotropic linear elastic material with the compliance S and loaded by bending moment $M$ and the axial load $P$." My head aches; I am in trigonometry class again, twisting, twisting slowly in a helical wind. These are the words of the tree nurturists Seubpong Leelavanichkul and Andrej Cherkaev, whose *Why Grain in Tree's Trunks Spiral* argues that the spirals are a compromise between the need to transport water to all parts of the plant, even those with no roots located directly beneath them, and the apparent fact that spirals angling greater than 37° weaken the trunk. Left-handed as I am, I cannot vouch for their math, but their title, *Why Grain in Tree's Trunks Spiral*, suggests grammatically that they studied but one's tree's multiple trunks.

Yet one need be neither mathematician nor grammarian to discern the inner longings of a tree. Many a lightning-struck trunk betrays its soul's tortured twisting by the split-bark course the lightning took—spiraling round and down, following, foresters think, the grain's helical twist. Tall trees draw lightning, and those exposed on the mountains ringing the Valley bear ample witness to the danger of standing out in a crowd. Some trees suffer more than others, those with much interior moisture most at risk. My son Edward and I have come upon tulip poplars literally exploded from the force of their internal waters boiling, crowns thrown to the ground, trunks jaggedly decapitated, fifteen-foot daggers of wood impaling the earth around them. Others' bark explodes in fragments, the lightning's track gouging the whiter inner wood. Such scars last a lifetime, scabbing over if not too deep or broad, trees wearing such marks more proudly than a nineteenth-century German did his dueling scar. Too deep, the wound never heals, inviting rot. But such strikes, healed or fatal, often as not twist round and round a tree and testify that even the gods' anger bends to the ways of trees.

# STUMPED

We have been long obsessed with stumps—stumps of limbs, of arms, of legs, of trees' trunks, all from the Germanic stump—"mutilated, blunt, dull." Look up the word on the Internet, alas, and you will find today's obsession lies in removing them. Forgive my stump speech, but I beg to differ. I would honor the stump. Look for the stump, it is ubiquitous, coming in all sizes, shapes, and ages, from tiny ones just small enough to stump your toe upon to the legendary ones Californians built dance halls on, something the poet Wallace Stevens transplants to the Midwest:

> In Oklahoma,
> Bonnie and Josie,
> Dressed in calico,
> Danced around a stump.
> They cried,
> "Ohoyaho,
> Ohoo" . . .

You can learn a lot from what at first may seem but rotting wood. And, learning, you can cry like the astronomer A. E. Douglass: "Ohoyaho, / Ohoo."

Douglass invented the scientific study of tree rings when he used those he found in old buildings in the American Southwest to trace the history of sunspots. Trees grow thinner rings in dry seasons, and Douglass used such annual changes to track the effect of sunspots upon Earth's climate—and to track the rise and fall of sunspots by the telltale trace of tree rings. Archeologists are forever grateful to him because his work permitted them to correlate and date the abandoned ruins of the region. Working backward from living trees' rings, Douglass and others sought to link these to the rings in beams they found in ancient pueblos. Correlating the "curves of growth" in ruin after ruin, they came upon HH-39, the Rosetta Stone of American archeology, which gave them a yearly chronology back to 700 A.D.

Sit yourself next to a stump and learn its history. Each year, the tree laid down new growth as its trunk widened. Each year, early growth is typically quicker than later growth. Being quicker and less dense, spring growth is lighter in color than the slower, denser, darker summer growth, providing to our eyes a succession of rings that give not only age but also growing conditions. Plenty of rain and favorable temperatures permit more growth and hence wider rings; drought and unfavorable temperatures reduce these widths. Even if you do not know the year the tree was cut, you can infer the climate it grew in. Which is what Douglass and his colleagues did.

Play dendrochronologist yourself. Chance upon a white pine just cut, stump here, trunk there, and sit and count the rings, one, two, three, four, and on and on until the day time stopped and the chain saw began. Observe the varying widths, the trace perhaps of damage to the tree. Pretend you are a scientist and write its history. Should the chain saw be yours, cut a two-inch-wide section of trunk, sand, shellac the surfaces, and you have a portable SparkNotes to the life of one *Pinus strobus*. Take this and read the same history in the trunk that lies still bleeding sap beside its stump. Each year, the white pine puts out a new whorl of branches, its trunk the visible record of its life, the distance between whorls as clear a sign of growing conditions as the width between its rings. Go and correlate. Here she or he or it was born, and here she suffered long years in the shade, and here the canopy was opened and she grew faster than you can count her rings and whorls, and here her tip was took and side branches fought to become the leader, and here one won, and here it left the others in its shade, and on and on until death came at your resin-coated hand.

Fell now a forest of white pines and match the records of trunk and stump to stump and trunk and read in the book of nature the handiwork of God, the sun, of man and global warming. It's intoxicating to learn such stuff, tempting to cut just one more tree. Which is what a graduate student, Donald Currey, did in 1964. Wanting to know the age of pine tree WPN-114, he cut it down. And so killed Prometheus, a 4,862-year-old bristlecone. The prophet Daniel spoke of such blasphemy when describing Nebuchadnezzar's madness: "The tree grew, and was strong, and the height thereof reached unto heaven, and the sight thereof to the end of all the earth: The leaves thereof were fair, and the fruit thereof much, and in it was meat for all: the beasts of the field had shadow under it, and the fowls of the heaven dwelt in the boughs thereof, and all flesh was fed of it. I saw in the visions of my head upon my bed, and, behold, a watcher and an holy one came down from heaven; He cried aloud, and said thus, Hew down the tree, and cut off his branches, shake off his leaves, and scatter his fruit: let the beasts get away from under it, and the fowls from his branches." After such knowledge, what forgiveness? Yet be not overhasty to cast cones at Currey; you read a book made of the flesh of pines. Send not for whom the saw sounds; it sounds for all, for bristlecone and white and yellow pine, for oak and hickory and poplar, for you, for me, for all that lives. Read then the living record carved in the bark of pine and in our wrinkled skin, the pine's green crown and our graying heads.

Pines die when you fell them. Most conifers do, the bald cypress and redwood two exceptions. But many a hardwood stump sprouts new life in the spring. When it is done deliberately, we call it "coppicing," a kind of renewable agriculture practiced on forests for millennia. Coppicing comes close to making trees immortal, the new sprouts on old stumps juveniles, and the British still coppice stumps eighteen feet across. Long-suffering Job too knew the miracle of coppicing: "For there is hope of a tree, if it be cut down, that it will sprout again, and that the tender branch thereof will not cease." The sprigs of trees, like hope, may spring eternal, "But man dieth, and wasteth away: yea, man giveth up the ghost, and where is he?"

Today trunks are carried off to sawmills, their bereft stumps weeping sap. As life must go on and forests be repopulated, the stumps dry their tears and coppice themselves, a new grove of trees rising, oddly grouped and multi-trunked. We walk through these survivors without remarking that their growth remembers slaughters decades gone. I read once that nearly every oak in our forests is coppiced and have since observed the legions that grow cheek by jowl, closer kin than my siblings or my son are to me, clones of themselves. "Genets," the biologists call them. The iris and day lilies you divide in your backyard are genets, each identical one to the other, the hedge of Leyland cypress your neighbor planted is a line of clones, the Callery pears in Walmart's parking lot all grown from the same plant. We but imitate nature. It's too far for me to drive to Utah's Wasatch Mountains, where Winnie the Pooh's Hundred Acre Wood is composed of forty-seven thousand clonal aspens, quaking root to root. Far too far to fly to Tasmania to see the five hundred clones of

King's Lomatia, the solitary clump of forty-three-thousand-year-old sterile trees that reproduce only when their fallen branches root. Closer to home, I can content myself with the heady perfume in spring of a clonal grove of flowering black locusts, the vibrant yellow of self-spreading forsythia, beg pardon of its clones when I dig the roots of a clump of sassafras sprouting along a fencerow, dodge the thorny thickets of identical honey locust and wild plum, and slide my hand across, around the twinned and tripled, quadrupled trunks of red oaks rising, like all creatures do, from their parents' ruin. As Isaiah promises, "There shall come forth a shoot from the stump of Jesse, and a branch shall grow out of his roots."

Stumps speak in mysterious ways to us. The dendrochronologist Champollions have deciphered one of their languages. Others talked in medico to Tom Sawyer, who knew you stick your warty hand in the waters of a hollow stump at midnight and chant, "Barley-corn, barley-corn, injun-meal shorts, / Spunk-water, spunk-water, swaller these warts." Less brave was the English poet Hugh Sykes Davies, who warned that "in the stumps of old trees where the hearts have rotted out there are deep holes and dank pools where the rain gathers, and if you ever put your hand down to see, you can wipe it in the sharp grass till it bleeds, but you'll never want to eat with it again." But what does Davies know? He calls the "dank pools at the bottom where the rain gathers and old leaves turn to lace" "a sodden bible written in the language of rooks." Are we not called to read the Bible? So go on, warty or not, plunge your hand into the watery word of God: it bathes in tea steeped in tree history. The trees are a many-tongued tribe, whispering in the winds that fan their second-growth leaves and needles their history, that of their parents, of the gone-forever forest primeval. Listen, and learn.

# OLD FIELDS

The tulip poplars in Petite's Gap rise impressively straight and tall toward heaven. And you think, here in this grove of two- and three-foot-thick wood titans, I am in what the red man knew before the white man came and cut it down—forest primeval.

Fat chance. These impressive trunks have risen in the three score and ten God grants mere mortals, their greatness testifying not to God or the federal government's preserving what God has given us but to the glory of the Appalachian cove. Scoop up a handful of the soil you stand upon, black, rich, damp even now in the summer's scorching heat. This is ground upon which tree or man may take his stand.

You see, but you do not observe, the insufferable Sherlock Holmes told long-suffering Watson when he confessed he did not know the number of steps leading to their rooms. Seventeen. Those to my upstairs? Fifteen. To my former office? Twenty-six. I am obsessed with numbering stairs ever since reading *A Scandal in Bohemia*. And observing trees ever since I learned that a grove of tulip poplars of uniform age could not possibly be ancient.

For an ancient forest in these parts will be a forest of diverse species, of divers ages, sizes. And the floor littered with decaying trunks. In this Baudelaireian "temple où de vivants piliers / Laissent parfois sortir de confuses paroles" (temple where living pillars / let sometimes issue confused words), we must learn again to read. So let us begin with forest phonics. Just as a planted pine plantation betrays in its uniformity its artificiality, so too a grove like Petite's suggests a forest sprung from a single season's sowing. Not that enterprising foresters raised this wooden temple. No, these pillars were sown by long-vanished tulip poplars, no doubt cut down by the loggers in the year before whoever owned this land sold it to the government. See, these conjoined behemoths here sprouted from the stump of some long-logged progenitor whose girth the seven feet between their hearts remembers. Around them rise single trees of lesser girth, sprung from seed or lesser saplings.

And, here half-hidden in leaf litter is a pile of rocks. And here, another. Cairns, raised to what obscure end? Depends whom you ask, whom you believe. Legend has it they are memory cairns raised by traveling Indians, who laid a rock here every time they crossed the gap. Yet others will that these are burial grounds, and beneath each cairn a warrior lies. The more prosaic think them where farmers threw the rocks that choked their fields. And, truth be, plenty an old cabin site's surrounded by the same. So, Sherlock, my inner Watson asks, which are they? Elementary, my dear Watson, elementary. We are in an old field gone to forest.

And so we are. These woods, which I'd like to think eternal, were cleared field eighty years ago. Over on Cole Mountain, you find rock walls running through woods for miles and wonder who in hell grazed cattle on such heights. But someone did, and where oak and hickory, maple and basswood, tulip and cucumber magnolia shade hikers today, yesterday cattle grazed in open sun. Wander the wildernesses the government has declared to grace these mountains and find in fallen wall and sunken cabin site the trace of tougher folk than we, who heaved thigh high the stones we stumble on to build their walls, whose stone chimneys, mortared but by gravity, still rise and still work, are you so inclined to try them. Lay your sleeping bag here on the loam filling this cabin gone so long there are no walls, and feed imagination as you feed a fire on a hearth three generations cold.

Slowly, glacially slowly by human standards, the ancient forests return. Tulip poplars love sun, and there are few saplings of their tribe now growing in the shade of Petite's Gap. When these great trees die in another three hundred years, God being more generous with trees' longevity than ours, beneath them will already be established a forest of the shade tolerant. Even now, I spy maples, a beech, a basswood. The forest's future diversity lies in the shadows, biding time.

This is old field succession. To see it, we must descend to lower realms, to the Valley where, already, the future arrives. Farm by farm, agriculture here succumbs to farmettes and subdivisions. Everyone wants a piece of paradise, and, carving ours out, we end it for another. The old curmudgeons who farmed these fields now sprouting For Sale signs have died, their children fled to towns and cities elsewhere. Sell the farmstead; make a killing—and who of us would do differently? When Daddy dies and they can find no one to lease the land, they let it go. Nature takes over, and soon fields groomed yearly by cow and bush hog bristle with annuals that are everywhere and come from nowhere—the same you wrestle in your lawn—purslane, dandelion, ragweed. Two to five years abandoned, they fill with stickweed, shoestring, asters, daisies, goldenrod, and Queen Anne's lace, the colorful wildflowers you see along the roadsides. Should no one buy the land, blackberries and wild sumac invade, multiflora rose making entry impossible save where a cedar forest springing up has shaded out the competition. Emboldened, wild cherry and black locust leave the fencerows to venture into what once was field. In twenty years, these rabid colonizers are joined by tulips, ash, and maples, whose helicoptering seeds blow in from adjacent woods. Heavier-seeded oak and hickory come later, carried by forgetful squirrels. A shade tolerant understory arises. Hemlock—should they still be here in a generation—and beech grow up.

This, like all such schema, is idealized. Where are the white pines that dot the mountain forests? The rhododendron and laurel? Basswood? One could object forever. And we have ignored fire and storm, which can turn a fifty-year-old forest into a bare field in hours. And topography, substrate, and orientation. The old logging road up out of Petite's Gap begins in a tulip poplar grove that faces north, sheltered from the sun and desiccating southwest summer winds. As it climbs, it winds its way west and south, leaving the cove forest where the trees reach skyward and the ground is green with herbs for a barer, drier world of oak and hickory, under whose boughs a thinner understory struggles to find shade and water. The green triumph of trillium and bellwort dies, the browner shades of more drought-tolerant lousewort rule. Here on these weathered, thin, and acid slopes flourish winterberry, azalea, and rhododendron. Dry out and thin soil enough, and the scrappy Virginia pine succeeds where others fail. Lift up your eyes unto the mountains, and trace their ridges in the evergreen of pines, the deeper, damper-soiled coves where tulip poplars green in spring, fall in autumn yellow, stand gray and bare in winter. Rhododendron and mountain laurel tonsure the mountains' foreheads where little else will grow, and on the thin soiled, rock solid summits, Table Mountain pine offer themselves to bonsaiing west winds.

Limestone-tolerating, shade-loathing cedars abhor these mountains, but black locusts, among the first to invade a field, abide. And climbing upward,

I spy a few etiolated older trees, nearly every one adorned with shelf fungus that foretells their deaths. These were old field pioneers, among the first to establish themselves in what was field and logged-out forest. Shade intolerant, they have no offspring, growing old and ignored among alien tribes of oak and hickory. Bark in tatters, branches broke, these locusts, not long for this world, demand attention be paid. Take your pocket knife, kneel, and incise on the hidden side of the fungus that killed them the initials of one you love. No one will know, no one will see, the tree will fall, and in the dirt that covers all, the name of one will linger longer than do your thoughts upon the death of forests.

So too the wild cherries, whose mouth-puckering fruit children and birds compete for, are dying, save in wind-thrown pockets where they can prosper for a moment. Should you come upon a berry-laden tree, take and eat in remembrance of the dead a handful of bitterness. The East Coast is littered with "Old Fields," what colonists called the abandoned sites of Indian villages and fields. William Bartram mourned 230 years ago when he came upon "yet visible monuments, or traces, of an ancient town, such as artificial mounts or terraces, squares and banks, encircling considerable areas" moldering into nothingness, fields where beauty berry bled in purple and pawpaw went unpicked. A 1751 map of Virginia puts "Shawno Fields" at the mouth of the South Branch of the Potomac River, in a place still known as Old Fields, from where you might mail a letter to yourself to remind you of what is gone: the forests, the Indians who inhabited them, the settlers who felled them, the mountain folk we ran off so we might call this wilderness. Much of the old will never return, the chestnut, elm, and hemlock gone, replaced by new trees of heaven, Princess trees, and Callery pears. "The old order changeth / yielding place to new / and God fulfills his will in many ways / Lest one great forest should corrupt our ways."

# CROWNS

Tree people call the tree tops "crowns," and, though devoted to the republic, I too shall honor monarchy. Trees shall be my books and teach me politic husbandry. And we will first to the white pine, whose piney pyramid typifies apical dominance, the upward growth of the leader at the expense of its sibling lateral shoots. *Primus inter pares,* as are all crowned heads within their families. And yet, uneasy lies the head that wears a crown, and pesky branches yearning to be great must be kept in place with hormones, especially one called "auxin." Down, down, the crowned leader cries, all must bow before my majesty. So that, all being well, the pine grows like a nursery child's pine, pyramidal, the leader basking in the sun, overtopping, overshadowing the rest.

Yet, listen. When the wind stirs their needles, the siblings sigh their discontent and shake the trunk that they, they, they must serve that which lords it over them. And then they pray for a leader-toppling tempest that hurls crowned heads to the ground. Or that the white pine weevil, the worm that flies in the howling storm, will in spring feed upon their leader, who shall weep resin at his fate, bend head, and die. And then the race is on, every lateral branch, freed from tyranny, racing to be great. What was a pyramid becomes a top-heavy thicket of conspiring, competing branches, each striving to outtop the other, until one reigns supreme over its lopsided kingdom and relegates in turn his defeated siblings to subservience.

"Scotobiology" (from the Greek *scotos,* dark), they call this study of the things of darkness, and in its gloomy academy we must matriculate. Those straight-columned tulip poplars in Petite's Gap all exhibit apical dominance. But you will occasionally come upon a tulip, its top topped, several branches reaching sunward, each yearning to be the sole possessor of a crown. Back of Kathy Ball's farmhouse stands one such monstrous, misshapen tree whose competing side branches, grown fat on sunlight over decades, have raised, like Babel's citizens, their towers toward heaven. Which has them blasted for impiety. Amid the rotting wreckage of branches thicker than I am tall rise sapling tulips, each straight, columnar, heaven's lesson learned, and auxin in control. Go then to nature, worm, and learn from her the vanity of vegetable ambition.

The woman who owned my house before me had the hemlock out back topped, as do many in these parts, for fear an ice storm might beat her to the job. Deprived of monarchy, the tree embraced democracy and raised as if cupping its hand a multistemmed Congress of competing branches toward the sun. Perfect for Edward's treehouse, in which he played president, the patriarch of paradise, and, seduced by the heady delights of heaven, monarch of the glen, lord of the land, and king of all that he surveyed—until an envious storm blew house and tree to ground, and my son stood among his dreams' wreck and learned, as Boethius had of Lady Fortune, that, to avoid the fall, one should avoid the rise.

Not all trees harbor monarchical ambitions. The oak, reputed king of trees, exhibits—sssh, don't tell it—"weak apical dominance," so that it grows nearly as fast out as it does up. Granted, the sunniest-dispositioned leader will usually dominate, much as, to draw a family parallel, an older, happier sibling might her younger, gloomier brother. But, overbearing older sister or not, such a tree—often as not an oak or maple—will have the fatter, rounder, jollier shape of childhood's lollipop trees. And forks galore, each calling, me, me, me, to foraging squirrel and child, who may, years from now, sigh in regret that she took this, not that, road to the sky.

Even these jolly giants must raise their leaves toward heaven's sun, and so their forks grow skyward, year by year abandoning the human child, who can but stand upon the ground and dream of reaching a fork ten, twenty feet above her grasp. Unless she live in the Deep South, where kinder, gentler live oaks bend their arcing branches earthward so that she can clamber up above her duller-witted parents to where squirrel and bird and dream have habitat. Here in the Appalachians, such trees are rare, and Edward must content himself with a butt boost from his earthbound dad to reach the first tier of aspiration. But then he's free as his imagination to climb up, up, up to where his perch overtops the forest splayed beneath him and live free, if only for a moment.

These more democratically inclined vegetables attract the wrath of heaven's king, especially on the mountain's ridges, where ice and wind tear leaf and limb and lightning rivens bark and trunk. Were tree sap blood, the rocks here would be redder than Shakespeare's history plays, these crowns more dinged and beat than any bauble the Plantagenet squabbled over. Stop, lift up your eyes toward these oaks' forking crowns and count in their broken branches' gnarled and snaking shapes the living record of a war of sky and tree. Among the leaf litter lie more limbs lopped than after Chancellorsville, with no marble marker here to read writ on "the limb of *Quercus rubra*." Lunch late in the day under these trees when the sun is lowering to the west, darkness seeping out of rock and shadow, and grow paranoid: these trees properly belong in the Wicked Witch of the West's weird woods. They creak, they groan; like Birnam's, they move and mask who knows what thoughts in their green shade. Scotobiologists all, they speak like Hamlet's Horatio, "of carnal, bloody, and unnatural acts, / Of accidental judgments, casual slaughters, / Of deaths put on by cunning and forced cause."

And these horrid histories we must memorize before nightfall. How here this branch broke and its rival siblings sought to outstrip each other in their rush to light, how here one wrapped itself around another in an embrace as familial and fatal as the serpent's. Here how two rubbed each other's bark until their fleshes merged. And here how this quiet, seeming unassuming sibling snuck past its fighting rivals to cheat them all of what they thought their birthright.

Look at their oaken feet sunk deep in the soil of Virginia, where as acorns they threw back the rocks that would have kept them buried, rocks that form now necklaces through which they thrust their ways lightward. The winds that howl here move faster round their ample girths, just as a rock troubles, speeds a stream's waters, so that no leaf mold hugs their root. Winter's icy burden on a crown grown lopsided and the southwest wind's prevailing breath have heaved this oak northeastward, the ground to windward humped but holding. And here, its roots failing, this tree lurched earthward with no human near to hear its world turned upside down, roots higher now than branches.

Yet not all who fall die, not immediately. True, the crunched and crumpled leaves on this dead monarch testify to her decapitation during spring or summer. But those fallen few who reck not His rod raise to their own glory lateral shoots upon the fall of apical tyrants. I know a storm-thrown sycamore still half-anchored in the soil out of whom a rod comes forth, out of whose root branches grow—laterals raised in praise toward the sun or god they worship. Regard the ravaged monarchs of the woods, how many, felled but living yet, throw once-overshaded lower branches heavenward, a row of siblings still attached to mother's trunk competing for who will be the one and only tree.

See this oak's peculiar shape, which trigonometry might plot: it rises tree like, curves horizontally as if it would not rise, then turns again abruptly skyward—its form the memory of some catastrophe that broke its back and would have killed it—perhaps this rotted shadow of a tree that fell upon it decades gone and broke but did not sever trunk from root. So that the leader shoot, now earth turned, died, and, once subordinate, this side sibling took in a wood-bending arc its place and rose, rose heavenward.

And some die. Like these, so gone only their ghosts still haunt these slopes, this nearly vanished line in leaves marking a trunk returned to the dust from which it rose, in whose rich earth a half a dozen lesser trees have taken root, their misshapen roots the memory of the trunk they first grew upon. And these odd hummocks scattered through the forest, each standing watch over an oblong shallow as if someone, set to burying a dozen corpses, had been scared off? These are the legacy of trees so long gone to dirt their only trace the hole their rotted roots left when ripped living from the ground, the hummock the soil they tossed in agony. Sit on one and meditate upon the all-too-similar hole your body too shall fill.

And leave forest for home and hearth. There we shall confront in terms we can't deny our internecine wars, spouse against spouse, parent against child, child against parent, child against child, a Hobbesian war of all against all— and me, aging and accursed progenitor, who must fall—and happy fall—to make a place in sun for son.

# ABANDONED BEDS

You don't need to look far to find the origin of the rocks in the cairns on Petite's Gap. You're on a mountain, after all: compare one of the smaller rocks to bedrock exposed in the nearby stream—they match. According to the geological map, they're both Precambrian, bits of the Pedlar Formation, igneous granodiorite and metamorphic granite-gneiss—rock born miles underground from molten liquids boiling up from even deeper down and cooling so slowly crystals grew large enough to spy with your bare eye. Granite, which the later birth of long-gone mountains tortured into striped black and white gneiss. The stuff's so hard it still builds mountains—which explains the Blue Ridge Mountains hereabouts. But if you drive back down Elk Creek to Arnold Valley, out and across the James River into the Shenandoah Valley, you're on a sweeter-soiled limestone layer cake devoured by millennia of rain. Here you are lower in elevation and higher in population—at least of people and cattle, though not trees, which you're more likely to find chockablock on the Blue Ridge. In Valley fields, you'll find rock cairns as well, where early farmers heaved rocks littering their fields. These are likely to be sandstone cobbles, what the natives call "river jacks" when found in the Maury River. And hard to explain in fields a hundred or more feet above the river, unless you believe in Noah's Flood—which the geologists over at Jerry Falwell's university in Lynchburg say they do. Rocks in their heads, is what rival rockmen at Washington and Lee University might say, preferring these to be leftovers from when the Maury River flowed a hundred feet higher. That's a hundred higher feet of solid stone, a lot of Virginian bedrock to wear away over a time longer than the wait for school to end. Still, it's what I think I believe as I trespass through fields and stumble over cairns sliding down the cliffs' edge toward the river. Two hundred years of farmers shifting stone from here to there have nigh obliterated what brought these cobbles to these fields, but

24

along the cliff's precipitous edge you can walk what looks like all the world to be dry creekbed. Which it is—the dessicated ancestor of Whistle Creek, gurgling now over limestone bedrock a quarter mile upstream. Keep heading upriver, and you'll find yourself descending, terrace by terrace, great grandparent to grandparent to parent to today's descendant creek and river. Fossil creekbeds, if something as evanescent as scattered stones can be a fossil, laid down so long before Noah built him his arky arky nobody can say just when. And carried here from mountains whose nigh gone carcasses House and North must be. Heft a cobble, lick it, pretend you taste river water that flowed to the Atlantic when beaches were a hundred miles inland of where they lie today, when Europe was one hundred, five hundred, a thousand miles closer. And who knows, perhaps the same water, evaporated, condensed, evaporated, condensed a hundred thousand times over, drunk by dinosaur and mammoth, Adam, Noah, Nero, George Washington, and John Wilkes Booth, still flows a hundred feet below.

Ghosts of today's streams haunt the land they shaped. Over on Buffalo Creek, I can trace through pastures the faint outlines of beds the creek abandoned for its current love, see by the marooned sycamores and fading levees where it used to hug the road, and, where the grass is greener, taller, lusher, guess its once wet course. Meandering, after the Meander River in today's Turkey, whose course wound so round itself the river's name became a noun, an adjective, a verb. Meander in meanders, and you are meandrous, meandrine, meandering. The Greeks carved stylized meanders into their architecture, harder angled and as enduring as the incised horseshoes of the Maury that have cut a channel through limestone cliffs so tall cadets at the Virginia Military Institute rappel them. But Buffalo Creek meanders through alluvium, stuff so loose her meanders migrate downstream, so that a farmer long lived enough can spy the creek's course shifting across his fields, and sycamores feel their trunks sliding seaward with the snaking serpent that first seduced them to dangle toes in a current built on sand. In one meander as loopy as an in-love coed's handwriting, with hearts dotting the i's and smiley faces replacing the o's, the Buffalo bit off its own horseshoe, marooned a quarter mile of creek in what they call an oxbow, silting in, home now to beaver and the county's greenest, grandest stand of watercress.

In some places, ghosts talk. Or grumble at having abandoned light and life for the dark, dank underworld. Sugar Creek looks like it just absconded, its dry, rockstrewn bed unmade. Its water philanders a half a mile away and underground, thinking I do not know. But I do know and have sat and listened to a soggy mattress squeaking twenty dirty feet below the grass. Only when rains reminiscent of Noah's come does Sugar, choked with remorse, return home. Then no one sleeps, because she roars in rage and threatens to wreck her bed and anyone foolhardy enough to have built upon her banks. Perhaps Whistle Creek's fossil beds once looked like Sugar's and she too returned occasionally to them, until she left them high and dry for farmers to dismantle.

The nearby Natural Bridge is remnant, geologists say, of such a disappearing stream, and Cedar Creek, which now flows 215 feet beneath a bridge, once secret paramour of a cave as long as the narrow gorge through which she runs. Violators of the seventh commandment are often found out, and, Cedar's secret boudoir collapsing round her, she was flushed out, her bed exposed. George Washington is said to have been the first to cast a stone upon her from the Bridge's height. No doubt a cobble from her long-abandoned proper bed, a cobble not unlike those the farmers cast in sinkholes above Whistle Creek, like that I pick up from her fossil bed and heft now. The temptation's there, isn't it? To do what time and river could not, to toss it down a hundred feet. Why not? You'll not be the first.

Karst, the geologists call such lands, most often limestone like that in the Shenandoah. Composed of carbonate rock, which the always slightly acid rain riddles like Swiss cheese, these are the rocks you've seen in pictures of China's pinnacled Li River, those on which French cave men painted at Lascaux twenty thousand years ago, those that in Florida today swallow people and houses, and those in Slovenia's Karst Plateau, which gave them their name. Their waters love abusing cartographers who, like the Bible, would have streams stick to the nice blue beds they draw for them. But karst creeks have hearts they'd rather follow, be that across the yard, over the hill, or underground. One such sinks into a jumble of rock on a farm three miles from town and disappears, only to rise a quarter of a mile away in another farmer's field and farm. And over here another emerges darkly from the cave it cleft. The Greeks called such a place Ploutonion, the Romans Avernus, and claimed its breath noxious, death dealing. But the mouths of Shenandoah caves are as sweet and inviting as a lover's, whispering, in August, in cool, seductive syllables, come, come away. Unsuspecting humans and crawfish come, forgetful of the warning of the gone-mad-from-breathing-stygian-breath Sybil, who cautioned Aeneas, "easy is the descent to Avernus: night and day the door of gloomy Dis stands open; but to recall one's steps and pass out to the upper air, this is the task, this the toil!" So too Sugar Creek chants in mystic syllables from far underground her warning to all who would take up their beds and walkabout: "if such a yearning, twice to swim the Stygian lake, twice to see black Tartarus—and if you are pleased to give rein to the mad endeavour, hear what must first be done. There lurks in a shady tree a bough, golden leaf and pliant stem, held consecrate to...." And then a cobble falls, and all is jumbled, garbled, and, for you and me, regretful, halfway to hell, return impossible.

# FUNGI

Should you, walking the woods, come upon bits of fallen wood stained green, as if a fairy or graffiti-mad hiker had an inexhaustable supply of Magic Marker felt tips, pick them up. You hold the track of either *Chlorociboria aeruginascens* or *C. aeruginosa*, close fungal relatives once famous in the art world. Mycelia of *Chlorociboria* are what stain the wood. You've seen such mycelia, most often white, threadlike mats ramifying through or over fungus-infected objects—dead trees, matted leaves, rotting oranges. Those of *Chlorociboria* are too small to appear threadlike, but in aggregate they stain wood a lovely shade some call blue-green, others indigo, yet others malachite. Most often found coloring easily collected twigs and branches, sometimes *Chlorociboria* will stain the hearts of broken logs, and then it's decision time: are you up to carrying a ten-foot beam two miles down a mountain trail so it might stand, greenly grinning, in your parlor corner? Artists have done more than this for centuries, using "green oak" in intricate parquetry and inlay. The fifteenth-century Italian master craftsman Fra Giovanni da Veroni used bits of the wood in his elaborate inlays of Leonardo da Vinci's drawings of fantastical polyhedrons, and with it his fellow Italian Antonio Barili decorated choir stalls in Tuscany. Three hundred years later, English woodworkers constructed veneer known as Tunbridge ware. You'll have to be a better craftsman than I am to make anything of the scraps of wood I settle for carrying the two miles to my car, but collect them I do, to sit upon my mantle. Taking one website's suggestion that I soak my bits and pieces in acetone (nail polish remover), I even extracted a jar of lethal liquid nearly pretty enough to drink that I must someday use to dye a t-shirt.

*Chlorociboria* devours wood, and so is a saprobe, an eater of the dead—one of us, for what are we but eaters of the dead? Devourers of dead animals, dead vegetables, dead fungi? Now I know my calling, saprobe, my nature,

28

saprophyte, my soul, saprobic, that which is me, saprogenic, resulting from things rotten. In nightmares I lie with John Brown a-moldering in the grave, shrouded in the fungus Jessica Mitford's *American Way of Death* assures awaits us all, embalmed or not. A greener funeral than that an ecofreak would wish upon me. Why might not I and *Chlorociboria* both be cut in little stars of green and inlayed in the night sky so all the world shall be in love with us and pay no worship to the garish sun? But that is Juliet who speaks, unaware that she and Romeo, like you and I, must to tombs and necrophilia.

We and saprobic fungi. Such as brown rot, which promises to eat the coffin the funeral parlor has convinced me I must buy to save my body for that great and awful day when death and hell deliver up the dead. Gone within months, the cynic say, and take me to graveyards where listing tombstones speak louder than coffin salesmen of the evanescence of the earthly. Brown rot was once called dry rot, the dryness referring to cured—hence dry— lumber as opposed to wet or fresh. *Serpula lacrymans*—the weeping serpent —is the virulent European form, notorious in the Eighteenth Century for ravaging Britain's wooden walls, her Royal Navy. Surveying the fleet in 1684, Samuel Pepys complained that "The planks were in many places perish'd to powder . . . and the ships' sides more disguised by patching . . . than has usually been seen upon the coming in of a Fleet after a Battle. . . , till I have with my own hands gather'd Toadstools growing in the most considerable of them, as big as my Fists." This rot reveals its presence by turning wood into brown, bricklike chunks, degrading the whitish cellulose within its cells and leaving brownish lignin behind. Since cellulose gives plant cell walls strength and structure, you can pick up a piece of brown rot wood and crush it easily to dust. As, the Bible reminds, can God us: "for dust thou art, and unto dust shalt thou return." Get you to the graveyard and tell the polished coffins of the dead: paint you an inch thick, to this you too must come, brick by brown brick. Shall we, then, sit here and drink a glass of Amontillado sherry?

For the love of God, Montresor, seal our catacomb. RIP. To escape brown rot, you would place your coffin inside a sarcophagus, a stately stone memorial as lasting as the pyramids? "He! he! he!—he! he! he!"—. "He! he! he!— he! he! he!"—go on, go on—and while you're at it, look it up in a dictionary: sarco-phagus, from *lithos sarkophagos*, the flesh-eating stone of ancient Greece, that which will eat your corpse. Whited sepulchres, which indeed appear beautiful outward but are within full of dead men's bones, and of all uncleanness self-consumed with loathing and remorse. The stench of decay, faint as the half-whiffed morning breath oozing from your sleeping loved one's half-parted lips, even now rising in your nostrils.

In fact, brown rot smells merely of dust of the cellulose of softwoods, pines and the like, that it prefers. To rid us of hardwoods—such as the maple coffin you might wish to seal someone in—God has provided white rot, which preferentially attacks brownish lignin, leaving behind white cellulose, the stuff of cell walls, and shrouding the dead in a collapsed, spongy, somewhat stringy mess, such as that I've seen oozing like exploded zits out of the rotting corpses of oaks on Thunder Ridge. An exquisitely named variation is the simultaneous white rotters, degrading both cellulose and lignin, leaving, like the Wicked Witch of the West, a puddle of disgust. And who has not suffered the company of such saprophilic rotters, capable of turning everything to mush and leaving you feeling filthy?

A bunch of rot, you say? No doubt, no doubt. But collect it, nonetheless. Green, brown, and white memento mori of the fate that waits us all, rich, poor, or rotter. Others actually propagate white rot, sticking hunks of new-cut maple and walnut in plastic bags or under sawdust and leaves to encourage mycelia to sneak throughout, spalding the wood with attractively colored blue-black zigzags of demarcation where competing fungi erected Maginot lines of resistance. Beware, though—left to the saprobes too long, your wood is useless. And wear a mask when cutting it—inhale too many fungal spores and your lungs might, like the wood, be spoiled.

Less vile and yet more common is charcoal mat fungus, which drapes the dead wood it attacks with a seeming charcoal coat. Suggesting perhaps cremation is the way to go to dust. Forest stumps and snags so draped can easily fool one into thinking a fire passed this way. But touch a piece of charcoaled wood and your fingers come away black. Charcoal mat fungi are charier at sharing coats with strangers. A well-burnt piece of wood will often already be turned halfway to briquets, as cracked and bricklike as brown rot. Charcoal matting looks more as if it had been worked over with a gouge than baked in an oven, the peaks and valleys of such stumps reminiscent of those impossibly elongated mountain spires that rose behind the fairy tale castles that filled our childhood books. But no Prince Charming will come to waken you or me when we sleep the long, long sleep of Sleeping Beauty, a many-colored fungal coat—green, brown, white, black—wrapped round our rotting bodies.

# 1.618

In 2011, thirteen-year-old Aidan Dwyer won an award from New York City's American Museum of Natural History for his science project, a PVC "tree" with solar-cell "leaves" spiraling around it in imitation of the way oak leaves spiral on a branch. Dwyer's tree produced 20 percent more electricity and collected two and a half more hours of sunlight during the day than a similar tree whose leaves were arrayed symmetrically. At the same time, MIT engineers were rearranging uniformly spaced solar mirrors at a site in Spain to imitate the pattern one sees in a sunflower's seed head to reduce shading and inefficient reflecting of sunlight.

Both projects lit upon turning each successive leaf or mirror at an angle of 137° with respect to its neighbor. 137°: the smaller of two angles made by dividing a circle's circumference into two so that the ratio of the longer to shorter arc is the same as the ratio of the circle's full circumference to the length of the longer arc. So divided, the arcs are an example of the mystic golden angle, the *sectio aurea* or *sectio divina*—the golden or divine mean, which fascinated Phidias, Plato, Euclid, Fibonacci, and Kepler. And which underlies the structure of Renaissance paintings and architecture, Nautilus shells, hurricanes, and spiral galaxies. And the humble snails and plants we pass on our walks. Mystic mumbo-jumbo? The divine secret of the universe? Let's take a hike.

News articles didn't bother telling me what species of oak Dwyer chose to copy when building his tree. Perhaps his was a platonic tree. Oak leaves exhibit alternate phyllotaxy—which is science talk for saying they grow in a spiral around a stem, with one leaf per node. Imagine you're God, looking down on a perfectly vertical oak; what you'll see is the leaves spiraling round the branch at a divine or genetically preset angle. In Dwyer's experiment, he is God, and the angle He chose 137°. Which is to say, were the oak branch a flat circle, the leaves would appear at 0°, 137°, 274°, 411°, and so on. What oakologists want to know is how many times do you have to run your finger round the branch until you come to a leaf positioned exactly over another? This, the phyllotactic number of oaks, is 2/5—two full turns and five leaves round (not counting the first) before your leaf sits atop another, an arrangement the Connecticut Botanical Society says cherries, tulip poplars, walnuts, hickories, and sweet gums also favor. Others favor other fractions: basswood and birch prefer 1/2; sedges 1/3; and willows 5/13. So let's do the math for the oaken 2/5: two full circles of 360° each equals 720°, divided by five leaves equals 144°. Dwyer's solar leaves were only 7° off, not bad for a seventh-grade science project.

But why these angles? The oak's seemingly arbitrary adoration of two and five, basswood's basking in one and two, willow's weirdness for five and thirteen? Enter Leonardo Fibonacci, a thirteenth-century Pisan who popularized European use of Arabic numbers and a sequence of numbers later named after him. Fibonacci numbers are the series 0,1,1,2,3,5,8,13,21,34, and so on, each successive number the sum of the previous two: 0+1=1; 1+1 = 2; 1+2 = 3; 2+3 = 5; 3+5 = 8; 5+8 = 13, and on, and on, and on. Higher Fibonacci numbers, divided by their immediate predecessors, closely approximate 1.618, the golden ratio. We're treading on *Da Vinci Code* toes here. And the math about as rigorous. Dwyer and MIT used 137°, the golden angle is 137.5°, and the golden ratio 1.618; the ratio between Fibonacci numbers varies from 2 (2/1) to 1.6 (8/5) to 987/610 (1.6180327868852).

Short of fudging mystic mathematics, I'm at an impasse. Let's assume, as do biologists and did Dwyer, that a leaf is a solar cell. Where might evolution arrange to place it so that it most efficiently captures sunlight in a canopy it must share with others of its kind? Back to those phyllotactic numbers earlier: oak leaves pop up every 144°, sedge leaves every 120°, willow leaves every 138.5°. It's all ballparky enough until you remember that the oak's mystic placement of leaves is 7° off the golden angle, so that after five leaves, you're 35° off where you're supposed to be, and your GPS is making warning noises. And why would the god of oaks decide to twirl around himself five times while the willow's went thirteen? Is it just a crap shoot?

Studying plant stem cells, the nineteenth-century German botanist Wilhelm Hofmeister argued that they preferentially formed "primordia" in regions farthest from older collections of stem cells. These new collections of cells concentrate the hormone auxin, which, seeking light, results in newer primordia moving down and out as the plant stem grows. Studying Hofmeister's trinity of directions—out, down, and away—the Harvard mathematician-biologist Scott Hotton came up with a formula that predicted that primordia formation would necessarily produce golden angle spirals and approximate a Fibonacci sequence.

It's difficult to discover the oak's 2/5 phyllotaxsis in the wild—and nigh impossible to discern a mathematics behind the tree's arrangement of branches. But pine cones provide clear examples of an affinity for Fibonacci. And a cone's scales are but tightly packed leaves modified to protect the seeds within. The Scotsman D'Arcy Thompson, whose 1917 *On Growth and Form* prefers mathematics to natural selection as an explanation for much in nature, explains, "The surface of a pine-cone shews a crowded assemblage of woody scales, close-packed and pressed together in such a way that each has a quadrangular, rhomboidal form. Each scale forms part of, and marks the intersection of, two linear series; these run upward in a spiral course, one in one direction and one in the other, and are called accordingly diadromous spirals." The Internet is chockablock with examples of these diadromous spirals, many of which seem fishily Fibonacci based. Donald Collins of Warren-Wilson College's Physics Department describes the cone of a pitch pine with "eight spirals bending to the left, and thirteen to the right," adding "8 and 13 are consecutive Fibonacci numbers. Several pine cones of the same species also exhibited the 8 and 13 spirals." Well, yes, but that "several" bothers me—did "several" others exhibit non-Fibonacci spirals?

Collins notes too that the spirals are logarithmic, growing farther apart as they spiral outward. An observation that Thompson says the English architect Sir Christopher Wren made regarding snail shells: "Wren not only conceived the spiral shell to be a sort of cone or pyramid coiled round a vertical axis, but also saw that the magnitude of the angle of the spire depended on the specific form of the shell." Thompson explains that such a shell "follows a skew path with respect to the axis of revolution, and the curve in space generated by any given point makes a constant angle to the axis of the enveloping cone, and partakes, therefore, of the character of a helix, as well as of a logarithmic spiral; it may be strictly entitled a helico-spiral." Pine cones, snail shells, and what else? Coneflowers, sunflowers, daisies—many of the *Compositae* seed heads also form easily recognizable helico-spirals. Pineapples, broccoli, cauliflower, and lettuce heads also seem to be better at math than I am.

Evidence that God is a mathematician? Or an accident of packaging? Have these various seed heads all landed upon a logarithmic spiral as the most efficient and equitable way to pack seeds in a roughly circular space? Peas grow in a pod—but a pod-shaped sunflower would be pendulously perilous. So its seeds spiral outward. Hofmeister's trinity of primordial needs—out, down, and away—adapted to seed heads gives you logarithmic spirals closely approximating the magic 1.618. Since .618 of a seed is useless, the numbers hover, but do not match, the golden.

# MAPLE LEAF DRIVE

When my engineering students profess themselves unable to find a topic worth writing on for freshman composition, I give them a maple leaf and ask for an hydraulic analysis of its veins. It works, often as not. The imaginative—or desperate—take what they know of sewer systems and apply it to the shriveling sugar maple leaf I gave them.

Sugar maple both because that's what was planted outside the building and because its geometry fascinates the kid trapped inside me who marveled in middle school, where, with protractor, compass, and straight edge, we drew the world around us. Here was a leaf, a hundred thousand leaves, each doing the same, and with no teacher to instruct them. Observe, Sherlock Holmes orders your inner Watson, a maple leaf. Its stalk—petiole in scientifickeeze—sprouts five veins, each at a 45° angle from the other, as if each leaf, shot through with red and orange, were the rising sun, its veins the sun's rays, and we Ra worshippers enthralled by a leaf. From each secondary vein depend alternating tertiary veins. These too join at 45° angles, though their unions are curved, not angular.

Random weirdness, I long thought. And then an engineering student, desperate for an A in freshman comp, took me, via his paper, to Lowe's, a sort of toys-for-boys store for grownups who never quite grew up—where we rummaged through the plumbing section and came up with: a "4-in Dia. 90-Degree PVC sewer Drain Sewer Tee" for $3.13, a "4-in Dia. 45-Degree PVC Sewer Drain Pipe Wye" for $5.12, and a "4-in Dia. PVC Sewer Drain Sanitary Tee" for $5.62—at $13.87, cheaper than a bought-online term paper and more original.

Those five 45°-angled sugar maple veins reminiscent of Japan's flag are but five 45-Degree PVC Sewer Drain Pipe Wyes glued together and capable of pumping xylem, phloem, whatever, up, down, through the municipal sewer—oops, I mean the maple leaf you hold. And those butt-curving 45s that join the secondaries, as sweet and curving as those in Lowe's 4-in Dia. Sewer Drain Sanitary Tee, a boxful of curvaceous PVCs angling for your ogling, and each butt but $5.62.

But, you ask—and here you must imagine yourself in the basement, faced with linking a vertically descending washing machine drain to the toilet pipe and both to the city sewer somewhere out there in the lawn—what or who determines when you use a 45° Wye angle, a 45° sanitary Tee, or a 90° Tee angle?

The city inspector's codebook and the Reynolds number, of course. Named after the British physicist Osborne Reynolds, who studied the transition from laminar to turbulent flow in pipes, the Reynolds number is the ratio of inertial to viscous forces in a system, sewer or leaf. $R_e = v Ð l / V$, where $v$ is the fluid velocity, Ð the density, V the viscosity, and $l$ a dimension—say the length or width of pipe or vein. The lower the Reynolds number, the more laminar, the higher, the more turbulent the flow. The formula may apply, of course, to more than sewers and leaves. Given a random marriage, compute its Reynolds number, assuming $v$ the velocity with which quarrels arise, Ð the density of the husband, V the viscosity of the marriage vows, and $l$ the number of years married. In a well-organized essay not to exceed five hundred words, discuss the changes necessary to reduce turbulence.

Higher math. But regard your basement pipes again: the smaller, pressurized water pipes sport 90° turns, the larger, gravity driven sewer pipes prefer 45°. Then, thinking about all that viscous maple syrup flowing hither and yon in your maple tree, you might ask, is there an alternative to sewer pipes to which I might compare my maple leaf? And there is—the highways of Canada and America, which bring the syrup to us, and to which mathematically minded engineers have compared leaf veins, with primary veins the vegetable version of the interstate or Trans-Canadian Highway, secondary veins state or provincial highways, and tertiary more local roads. (Yes, I too thought only poets used metaphors, but apparently better-paid botanizing engineers do as well.) And, just as you compute alternative routes to the maple syrup superstore when you find a traffic jam, so too do leaves when they discover a caterpillar has just bitten Maple Leaf Drive in two. Mother Nature and the Highway Department have both devised multiple ways of getting maple sap to where you want it—the roots, the leaves, your kitchen table.

The Rockefeller University physicists Marcelo Magnasco and Eleni Katifori's experiments on leaves with fluorescent dye and a hole punch made NPR back in 2010. The two punched holes in leaves and then watched how the glow-in-dark dye they force-fed the leaves made it around the holes. Which it did by using the interconnected vein system you can see for yourself by studying the lacework in the well-rotted leaf you fished out of a stump. What remains are the hollow struts of veins, which provided both transport and structure to the living leaf. And they connect and interconnect repeatedly, so that, if Primary Interleaf Highway 1 is cut, the sap—which could be me or tree—can take Alternate Route A, B, or C. A sort of built-in Google Map for leaves.

Loopy—which is what Magnasco and Katifori call the system, cloverloop upon cloverloop. Loops that serve not only as alternative routes in caterpillar disasters but also manage fluctuating demands within a leaf. Say leaf 7684's right top and bottom side are in the sun and therefore need more water to offset their gaping stomata, but fat bully leaf 4392 overhead has shaded 7684 halfway up. How do tree-you get water to where you need it without shipping it through 7684's shut down and water-drenched sections? Use the redundant loops built into the system. Not that every leaf's on board with the idea of more is better. When Magnasco and Katifori holepunched a ginko leaf, the area beyond the hole never received dye. Such parsimony, however, did not doom the ginko to being eaten to death by dinosaurs. It's still around (although it didn't exactly prosper until humans decided to line their streets with it).

But why the maple's five primary veins? And why those floppy nothings in between? First off, those floppy nothings in between that look like they might tear at any moment are not called floppy nothings in between. They are sinuses, like those in your head. And, according to a trio of French scientists, they're there because of origami. Deciduous trees' next year's leaves overwinter folded up in buds. Applying the principles of Japanese paper folding to trees, Etienne Couturier, Sylvan Courrech du Pont, and Stephane Duoady discovered that the floppy nothings in between, aka sinuses, are where the leaf folds upon itself and, necessarily, cannot therefore have a rigid vein. Since newborn leaves are but miniature versions of grownup leaves, the sinuses remain. And the center of the sinuses, where the floppy secondary veins join, is where the leaf folded upon itself when in the bud. Different leaves fold different ways, and the Frenchmen take us through a fascinating botanical tour that culminates with their absurdly complicated but oddly beautiful rendering of a red cabbage as a form of botanical packaging. Because, they argue, a leaf's shape is but the accidental by-product of "a geometrical problem of packing" to avoid water loss over winter.

And there you have it: the maple leaf, a sewer system, the Trans-Canadian Highway, origami. Something to ponder when pouring syrup tomorrow morning.

# SHIT

Imagine dying for shit, losing your coastline for shit, international politics compromised over a hundred years for shit, and, shit, you've conjured the Guano War involving Bolivia, Chile, and Peru from 1879 to 1883, when the three neighbors clashed over who would control immense deposits of bird shit. Bolivia to this day resents the loss of her coast to Chile. And all for shit.

Polite term "guano," which word we get from the natives there who've collected it for more than a millennium. Banks 150 feet thick accumulated on dry coastal islands off Peru from birds depositing their riches. At first valuable as fertilizer, it became even more desirable when used to manufacture explosives. Rich in nitrate, fertilizers can be explosive, as were the two tons of ammonium nitrate Timothy McVeigh used in the Oklahoma City bombing in 1995. Today, both fertilizers and explosives use nitrates manufactured in factories, and the islands languish. A number of which you and I own, thanks to the Guano Islands Act of 1856, which permits an American citizen to claim "any island, rock, or key, not within the lawful jurisdiction of any other Government, and not occupied by the citizens of any other Government." Which is how we came to own various specks of land around the globe, uninhabited save by birds and their guano—some of which, like Midway Island in the Pacific, became famous in their own right.

Your car windows collect guano, too, though at rates, you hope, far below the commercially viable. Typically, bird droppings consist of a solid dark bit of feces inside a white paste of uric acid, which is fairly insoluble in water. Since most birds excrete (and mate and lay eggs) via one orifice, the cloaca, they piss and shit simultaneously. What would be urine in you or me contains significantly less water in a bird and so comes out as a paste, rather than as pee.

Pause before flicking. Not only does that spot of bird poop reveal the anatomical wonders of its originator and the often base origins of human warfare, but it is also used in Japan, I shit you not, for beauty facials. Apparently, the white makeup geisha and Kabuki actors wear once contained noxious minerals such as lead and zinc. Bird droppings were discovered—by whom is lost in the vapors of history—to restore the poison-ravaged faces. For $32, you can buy 1.6 ounces of bird shit from Amazon.com. Or you can scrape it for free off your windshield. In an example of no-need-to-translate, the beauty treatment's known in Japanese as *Uguisu no fun.*

Birds poop in woods as well as on cars. And you can glean much from examining their deposits before smearing them on your face. For years, buzzards have wintered in the woods behind Lexington's colleges from roughly Thanksgiving to mid-April. Despite efforts to eradicate them, they still flock here, drawn by what? our more-genial-than-farther-north climate? the carcasses of skunk and cow? the corpse of the Confederacy? The copious defecations in their roost have turned the trees—white pines—white, their needles, branches, trunks, and the ground beneath a carpet of noxious smells chockablock with bits and pieces of dinner and fallen feathers—making a vacation to the South American guano isles unnecessary and suggesting that the communal habits of dinosaurs must have led to acres knee-deep in such waste.

Buzzard lovers—and there are such—aver that this "whitewash," as they prefer to call what the less discrete might term "diarrheatic shit," is not the unsanitary horror it seems. Indeed, vulture love leads them to claim "vulture poop is actually a sanitizer. Amazing! Their stomachs contain digestive acids that kill virtually all bacteria and viruses, and there is even evidence for the claim that they can consume meat infected with anthrax, destroying the virus in their digestive system." No one—to my knowledge—has yet packaged vulture poop as a facial cream—although, to cool themselves off, buzzards are known to coat their legs with it. Which polite ornithologists call "urohydrosis," the evaporation of their liquidy whitewash doing what sweat does for us. As well as, assuming its excrement as hygenic as aficionados claim, killing whatever nasties they stepped in.

Buzzards are infamous for vomiting when startled, heaving up gobbits of half-digested, half-rotted roadkill to distract us while, lightened, they take off. Nor are they the only bird to throw up on a regular basis. Hawks do so, as well. So necessary to their health is such behavior that it's encouraged in falconry—falconers force-feeding their charges nuggets of feather or cotton to encourage such cleansing of the crop. Owls too are famous for their pellets, tight bundles of fur and bone and other indigestibles that they regurgitate. And which you can buy for three dollars a pop online. Which science teachers do, so their pupils might dissect the pellets and learn what Hooty dined on last night. Since owls prefer to vomit in the same spot night after night, if you find such a vomitorium, you will be long supplied of pellets. Edward and I, having come upon one in our wanderings, collected a Baggie's worth. But be forewarned: before examining the dinners of such birds, microwave the pellet to kill whatever Hanta virus may still be lurking among the bone and fur of the dead.

In the Shenandoah Valley, bird shit has become a problem thanks to your and my love of chicken and turkey. Commercial poultry farms—long, low buildings you pass by with increasing frequency on the interstate—produce tons of guano. Ohio State University estimates your average laying chicken produces three to four ounces a day, your average broiler perhaps four, and a four-week-old tom turkey five times that. Which doesn't sound like a heck of a lot, until you realize some large operations house 100,000 birds. West Virginia's eastern counties alone produce more than 155,000 tons of waste a year. What to do with all this birdshit's a problem. Enterprising hay farmers have teamed up with commercial poultry farmers and scatter the stuff on their fields as fertilizer. You can, in some places, buy bags for your own backyard garden. Should you live downwind of a commercial poultry farm, as does my friend Yolanda, take heart: that eye-burning, nose-wrinkling acrid stench wafting your way contains ammonia, which will fall to your yard, providing free nitrogen for your plants.

Mammal poop does the same. Take a stroll through a Valley field and you'll notice islands of lush green grass rising above the lower, browner sea of grazed grass. These are bovine equivalents of South America's guano islands, the grass growing lusher, greener upon an otherwise forgotten cow paddy, ungrazed thanks to cattle's instinctive avoidance of such poop-powered paradises, replete with nutrient and disease. Plagued with piles of poop, cattle farmers have resorted to recycling it, and many a dairy farmer, feed lots hock-high in shit, spreads his fields in spring with manure as odoriferous as avian poop. I myself have collected dried cow paddies to fertilize my garden. And Lacie has long used horse shit to enrich hers. On slow days, we debate the relative merits of avian (rich but liable to burn), cow (filled with weed seeds from grazing), and horse (filled with grains from feed) poop. We could resort to commercial fertilizer, but that's not as fun as shoveling shit in and then out of the back of a pickup.

Nor are we and our factory-farmed food the only animals to pile up excrement in noxious amounts. Heralding spring throughout much of America, silken tents in the crotches of cherry and apple trees are the homes of *Malacosoma americanum*, the Eastern tent caterpillar. Hatching from overwintering egg masses—blackish dried-foam wrapped round the tips of twigs—the colorful caterpillars shelter nights in their ever-larger tent, which they leave to feed during the day, laying down silken trails doused with pheromones, so that popular feeding sites are connected to the home tent by ever-larger silken highways that can stretch throughout a tree or, upon occasion, from tree to tree. Tents white as purity when first spun soon darken with frass and are a noxious nastiness should you try pulling them by hand from out a tree.

Their fall counterpart is the webworm, *Hyphantria cunea*, whose caterpillars feed within the tents they weave on the ends of branches. Like the tent caterpillar, the fall webworm fouls its nest, and you can judge one's age by both its size and that of the frass accumulated within. Less specialized eaters than tent caterpillars, webworms have been found feeding on more than six hundred kinds of trees. Both species thrash about if disturbed, and the movement's contagious, one jerking caterpillar soon joined by its siblings, so that a tent can become a writhing mass of inch-long terrors. Both can defoliate an entire tree, and last spring leafless wild cherries lined Rockbridge fencerows. The tent caterpillars threatening my apples had to go, Edward and I descending on them on a cloudy day, when they were all holed up inside, ripping their silken shrouds and piling them on the ground where we set them afire. Silk burns easily, and so we take burning rags tied to sticks and torch the too-high-in-a-tree's-branches-to-reach-by-hand webworm citadels, they and their silken ramparts blazing brightly in the evening's growing dark.

Smaller and less devastating are the myriad leafminers whose frass-filled tunnels decorate many a forest and garden leaf. You've probably come across *Pegomya hyoscyami* in your beet, chard, or spinach leaves. This fly's maggots excavate threadlike tunnels that snake between the surfaces of leaves, gradually enlarging until blotchy patches of chlorophyll-less brown spread through a leaf—the size of tunnel or blotch a quick way to guesstimate the age of your invader. You might not think pine or arborvitae needles edible, but leafmin-ers do, munching contentedly away on hedge and grove. While scientists can identify a leafminer by its tunnel's shape, less perceptive souls such as myself must content themselves with watching entire trees turn brown, which Vir-ginia's black locusts (*Robinia pseudoacacia*) do every summer. When miles of locusts seem dying midsummer along the interstate, it's time to take to the fencerows and examine the locusts' blistered leaves for the track of *Odon-tota dorsalis*. Their numbers are legion, each locust leaflet home to a hungry larva, each locust host to thousands of leaflets, and thousands of locusts lin-ing every Valley fence and road. Year after year they consume the leaves' in-ner green, leaving a skeletonized and dying exterior filled with frass. Which doesn't appear to harm the long-suffering trees, who bear with patience their yearly consummation.

Easier to examine is the excrement of larger beasts, the scat of deer, pos-sum, fox, and coyote. Horse, cattle, and deer are all herbivores, and herbivore poop tends to drop in pellets. Deer pellets resemble oversized sheep pellets, if you've seen those. If not, think of the road apples of horses, reduce, condense, and darken, and you have deer pellets, which may or may not remain in the clump they came out in. Clumping, some think, depends on what the deer's been eating—moister berry and apple poop clumping more readily than drier leaf and twig browse. Hard enough to pick up and examine for the tiny traces of what their depositor has been eating—grasses, twigs, leaves, berries. Size may or may not matter: deerologists desperate to publish toilet papers have de-termined that the average deer shits thirteen times a day, that the average shit holds seventy-five pellets, and that bucks and does, like humans, leave equally large loads. Should you think yourself above rolling deer pellets between your fingers, consider that dedicated deer hunters have been known to smear their clothes with them in order to fool deer into thinking them ambulatory piles of shit. And that rabbits regularly re-eat their pellets in order to obtain nutrients they missed the first time through. These coprophagous rodents leave round, slightly squashed pellets of only vegetable material, whose color varies both by freshness and food source. But always dry enough to handle.

Just as you can tell when you last ate corn, bean, or beets, so too you can guess the dinner menu of the poop you find upon the trail. Possums adore pawpaws and persimmons, and fall is a good time to come upon possum poop as filled with seeds as are persimmons and pawpaws. And a visual explanation for how these trees manage to move from place to place. So too the berry-seed–filled poop of raccoons, which may also have its share of crustacean shells—fiddler crabs on the coast, crawdads inland. A coon's scat is generally Tootsie Roll–shaped, blunt on both ends, while a possum's is more variable and uneven. Bears too eat berries, and their large droppings can be richly stuffed with seeds and dyed the color of blackberry, blueberry, or raspberry. Or filled with ants, which they adore and which they search for by ripping up the rotting trunks and branches they come upon upon the trails they share with us.

Black bears are primarily herbivores, despite their sometimes grizzly reputation, and so their poop pelletizes. Carnivore shit, however, sticks together, usually in cylinders with a tail at one end. You can run an experiment on yourself, if you're willing to look in the toilet before flushing (perhaps in more ways than one) and correlate consistency and shape with diet. To pretend to be a herbivore, abstain from meat for several days; to pretend to be a carnivore, eat meat. And then regale your friends with the results. So that, should you find a cylindrical turd graced with a tail upon the trail, you'll know you're dealing with a carnivore. And, if so inclined, you can dissect the various leavings you discover and learn whether meat was on the menu, coyote scat often thick with fur and bone too crushed to determine its species. Or, coyotes being as omnivorous and sweet toothed as you and me, blued, reddened, or blackened with berries.

Size matters in matters fecal. Fox and coyote scat is quite similar, but fox is smaller—unless the coyote a runt and the fox a brute. And, if you cannot really tell how big the animal was that left its calling card upon your path, you can accurately gauge the size of its anus by the diameter of its fresh scat. So bring a ruler on your rambles. Wash it and your hands afterwards.

P.S. (Latin for polite shit): One invasive alien mammal's scat, increasingly common on trails, is ours. Some *Homo sapiens* seem reluctant to leave the trail to do their doo, and you come upon their trace with depressing regularity, piles of shit and rotting toilet paper festooned where some hungry beast dragged it. Should you, like Pooh, desire to poop in the woods, walk off the trail a decent ways—the Appalachian Trail Conservancy suggests eighty steps—dig a small "cat" hole six inches deep and wide with knife or trowel, defecate, and bury. Do not, they counsel, bury under a rock: that just delays decay.

# THE VEGETABLE GOSPEL

🌱 Shakespeare's Duke in *As You Like It* proclaims that "this our life, exempt from public haunt, / Finds tongues in trees, books in the running brooks, / Sermons in stones, and good in everything." Which one might take for the motto of this chapter, in which the gospel according to vegetables is discovered to be writ large upon our forests' plants. Were you, like me, to go truant from church on Sundays, though wandering in the wilderness, you still will have the straight and narrow preached by the plants beside you.

Scoff, will ye of little faith? Observe, then, the dogwood, which blooms in time to make it to Easter services. Once upon a time, some say, as big and strong as an oak; they crucified Christ upon it, and he, to comfort the tree, turned it so small and weak it could not again be used to hang someone. He ordained its flowers, white as angels, Easter blooms, their shape cruciform, stained with dried blood, and in their center a crown of thorns that turned to blood red drops in fall to remind the world of His sacrifice. Amen. Or so I learned in Sunday school, too young then to wonder how the New World's *Cornus florida* made it back to the Old in time for Good Friday. Perhaps renegade New World Lamanites sold *Cornus* logs to the Romans following a hitherto unknown pre-Columbian trade route. Or they remembered the reed Ra boats of Egypt and floated, logs and all, back to Egypt. Skeptic science may pretend the conspicuous flowers are but bracts, modified leaves, the crown of thorns a cluster of tiny bisexual flowers and the fall fruit not blood red drops but drupes, their color a fruit flag waved for hungry birds. But the faithful know better.

Which goes as well for the redbud (*Cercis canadensis*), blooming with the dogwood. This is the tree upon which Judas hanged himself in shame for having betrayed Christ. And where I came from, that's what we called it, the Judas tree. Dismayed by Judas's misuse of its boughs, the once mighty redbud shrank, its flowers blushing to this day in shame, its leaves heart-shaped in their love of Christ. Judas, of course, didn't use a North American redbud; he had the *Cercis siliquastrum*, native to the eastern Mediterranean and which people over there call the Judas tree. Though, skeptics aver, that may be a corruption of the French name, *arbre de Judée*, tree of Judea. Judas or Judean, the tree bears flowers both beautiful and delicious, well worth gathering to glorify a spring salad.

A happier plant, the Christmas poinsettia (*Euphorbia pulcherrima*) prefers proclaiming Christ's birth. Various versions of its legend involve a poor child with no gift for the baby Jesus. She, or he, or they pick some weeds to lay in the creche, and they miraculously turn into a bright red starburst of a flower, recalling the star of Bethlehem, in whose center is the golden crown of Christ the king, its bloody color reminding us of Christ's sacrifice. And, indeed, in Spain it's called *"flor de Pascua"* or Easter flower. Of course, science yet again feels the need to point out that the flower's red petals are actually bracts, not petals, the flowers being the small yellow objects making up the purported crown. Nor will you find this native of Mexico in our woods. But in the Deep South you may stumble, as I have, across its close but paler cousin, *Euphorbia heterophylla*, leaves touched with red as if it had wanted to become a poinsettia but changed its mind. All *Euphorbia* reduce their flowers, and a magnifying glass reveals them to be cup-shaped—more easily seen in *Euphorbia cyparissias*, or cypress spurge, which grows in waste areas through much of America but, as far as I can tell, preaches no gospel. And for those made euphoric by *Euphorbia*'s gospel, beware: the white sap oozing from its leaves and stems can blind you—though only temporarily. Upon your cure, you can sing with John Newton, "I once was lost, but now am found, / Was blind but now I see."

With the passion flower (*Passiflora* ssp.), God outdid Himself, packing in it Christ's crown of thorns, the Crucifixion's three nails, the scourge with which Christ was beaten, and the column to which he was tied when beaten, at the base of which are five stains representing the five wounds. The purple-tinged petals recall the shape of the centurion's spear, and their underside is spotted with Judas's thirty pieces of silver. The seed pod, sweet and delicious, is the world Christ came to save. Jesuit missionaries are said to have converted Indians using the passion flower as their text. And who could resist devouring such a complicated vegetable gospel. If your only contact with *Passiflora* has been passion flower juice or tea, get thee to the highways and hedges, where *Passiflora incarnata,* the maypop or purple passionflower, preaches passionately. But if, like me, you should be inclined to take it home with you, be forewarned—the maypop took Christ at his word when "he said unto them, Go ye into all the world, and preach the gospel to every creature." It will not be confined, spreading itself throughout your flowerbeds and covering your trellis with its leaves.

God also left his gospel text imprinted on the holly (*Ilex* ssp.), whose leaves, like Christ and Christians, never die, whose thorny edges recall Christ's persecution and crown of thorns, whose red berries remember the blood He shed. Should you pooh-pooh such tales as superstitious nonsense, you've good company with the Church of the Great God, which excoriates those who observe Christmas as participating in "sheer paganism directly descended from ancient rites practiced long before Jesus' birth." Which reminds me of my mother's rejecting Santa and stockings as pagan when she became born again. 'Twas the night before Christmas and no one had dared hang by the chimney a stocking with care. But Christmas is the season of miracles, and Christmas morning a solitary stocking hung—filled with gifts for Mother. The next Christmas, a stocking hung for everyone.

Lynnaeus, the father of binomial nomenclature, also read the gospel in America's jewelweeds. We have two in Virginia, yellow and orange, both *Impatiens,* whose cousins you have probably bought in spring to plant as annuals. All are *Impatiens* because, impatient, they fling their seed, when ripe, at the slightest touch—an example of explosive dehiscence you can test in your backyard or on your walks. Linnaeus, whose sense of humor a number of plants' Latin names reflect, named one *Impatiens noli tangere.* Scientists untutored in the gospel may suppose *noli tangere* but Latin for "don't touch." But the devout hear an echo of the risen Christ's warning words to Mary Magdalene in John's gospel, "Touch me not," in Latin *"Noli me tangere."*

# "WHAT PLACE IS THIS? WHERE ARE WE NOW?"

From February 21 to December 18, 1916, in hilly terrain north of the French city of *Verdun-sur-Meuse*, 976,000 French and German soldiers died, 97,600 men each month, 3,253 a day—though nobody really knows the number. Nearly half of the dead have never been found, most blown to mush by 44 million rounds of artillery shells. Among the monuments is la Tranchée des Baionettes (the Trench of Bayonets), which remembers France's 137th Infantry Regiment's third company, annihilated during the German bombardment and twelve of whose bayonets were found sticking through the dirt that had entombed men who refused to run. Or not. History, as always, is more equivocal. Some say gas killed the men. Others that they died in a German charge. Yet others that the bayonets were affixed after the battle to mark the graves of twelve men. Like the number of dead, nobody knows. But the story of glory survives, and, surviving, seduces other young men into dying gory deaths in needless wars.

Virginia too has her share of battlefields—the Revolutionary Yorktown and six national parks where were fought the two battles of Manassas, the battles of Fredericksburg, Chancellorsville, the Wilderness and Spotsylvania, Seven Days, Cedar Creek, Petersburg, and Appomattox. I have walked Manassas's rolling hills, stood on Henry Hill and then strode down to the clay rut where my ancestor Henry Augustus Middleton took a Yankee miniball to the lungs and spent a week dying, coughing blood out in some corner of a field far, far from home. Sundays my mother wore to church a locket with Henry's hair crossed over that of his cousin Oliver, who died at Matadequin Creek during the battle of the Wilderness. Yet the Manassas clay I've held was no redder than that in my back yard. And neither Henry nor Oliver was important enough to merit memorial more than a locket now gathering dust in my sister's bureau.

Rockbridge County also has had its share of battles. Four miles west of Lexington, a bronze plaque bolted to a rock remembers October 10, 1764, when the Shawnee chief Cornstalk and his warriors killed "fifty to sixty persons" near Big Spring. Assuming the stone is accurate. Oren Morton's *History of Rockbridge County* claims two attacks took place on Kerr's Creek, one, on October 10, 1759, in which twelve settlers died and thirteen were carried off and another on July 17, 1763, when, depending upon whom you believe, seventeen people died, or perhaps sixty, or maybe eighty, and eleven, or twenty-five, or thirty were carried off. Again, nobody's quite sure. And nobody talks much about these dead men, women, and children. If it weren't for the plaque today's cars whiz by, I'd have never heard of the Kerr's Creek massacre. Or massacres. Whichever.

A quarter of a mile away, the McKee Cemetery occupies a sloping hill overlooking Big Spring, where I stop to rest on bike rides. The graveyard being large, the store-bought grave stones oddly placed, locals believe the massacred are buried here, unmarked save for an occasional field stone. The names we know include the Cunninghams, the Doughertys, the Gilmores, the Hamiltons, McKee, McMurty, Ramsey, Stephenson, Thompson, Wilson, and Winyard. The willows weeping into the pond at Big Spring are those a Kerr's Creek settler, Jenny Gilmore, conjured when, captive and told to sing a hymn, she sang Francis Rouse's Puritan translation of psalm 137:

> On Babel's stream we sat and wept,
> When Zion we thought on,
> In midst therof we hanged our harps,
> The willow trees thereon.

What Lexington's Virginia Military Institute thinks on is the Civil War battle of New Market, where ten cadets died fighting seventy-five miles to the north. The school raised to their memory a monument forty years after the war when both sides, busy honoring a generation of veterans then dying off, erected two armies' worth of stone and bronze soldiers, Northern facing south, Southern facing north, ready to square off again even as the graying veterans were shaking hands. You'd think, from the balleyhoo the school plays out each May 15,

they'd have gotten right the number of boys dead. But no, they missed two—
as small bronze plaques correcting the mistake beneath the statue of Virginia
Mourning Her Dead make clear. Like Verdun, one supposes there were just
too many to keep track of.

Outside town, the county high school backs up to Hunter Hill, and my
son Edward's cross-country team runs the woods that stand where the Union
general David Hunter set his artillery on high ground overlooking the Maury
River, shelling VMI and driving the Confederates out. The path winds through
a grove of large white pine, flanked by larger oaks, none old enough to have
witnessed Hunter's cannonade or what he described as the returning fire of
Southern "sharpshooters posted among the rocks and thickets of the opposite cliffs" near where VMI's rifle range now echoes noisily with tomorrow's
would-be snipers. The Institute's website announces that "There were no ci-
vilian or military casualties as a result of this occupation," and Hunter himself
mentioned none in his reports. But the Ohioan J. O. Humphrey's diary notes,
"Geo. W. Tank of our company was killed, no other loss was sustained except
three killed and four wounded." And sixteen-year-old Lexingtonian Fannie
Wilson's letter tells us, "The Yankees took Mr. Matthew White, Jr. prisoner
and he was seen Sunday afternoon marching out of town with a squad of
soldiers, who shot him for bush-whacking; all the time deceiving his parents
by telling them he was at home. His body was found unburied in the woods
near Mrs. Cameron's house on the evening the Yankees left." Edward and his
teammates cannot know, because I do not tell them, and Lexington has cho-
sen to forget that George and Matthew, boys about their age, and three others
no one thought worth the naming, died here in summer long ago.

Stuff like this drives me to the mountains. Six trafficky miles by bicycle
to Buena Vista—named after a Mexican War battle—and then four tough
ones up to the Blue Ridge Parkway. But, 1,500 feet above the Valley, war and
death seem distant. A mile north is White's Gap Overlook, where John Jordan's old toll road crosses. When I first came to Lexington thirty years ago,
you could still drive the trail. But the Forest Service has since barred it, and I
have to fight scrubby pines for a right of way when I come down on my mountain bike, bouncing over Antietam quartzite, named after stones dyed red on
September 17, 1862, with the blood of 22,000 Americans. Here, the quartzite
cliffs are ribbed with *Skolithos linearis* worm burrows dug in the sands of vanished oceans 500 million years ago.

A mere 140 years ago, the Union general Alfred Duffie pursued retreating Confederates up this same road, "[d]riving a small force from White's Gap . . . without serious loss." A Frenchman, Duffie knew serious loss—having seen action at the Battle of Balaclava in 1854, famous today because of Tennyson's *Charge of the Light Brigade,* which asks of the men who died there, "When can their glory fade?" Perhaps theirs never can. But in Virginia, Jordan's road fades back to forest, and, reading the sign at White's Gap, you know only that you are at 2,567 feet. Because no one remembers the names or number of those who died here in a long-gone June.

# THE DOCTRINE OF SIGNATURES

The English Puritan Richard Baxter wrote that "The world is God's book, which he set man at first to read; and every creature is a Letter, or Syllable, or Word, or Sentence, more or less, declaring the name and will of God." And you and I but kindergarteners, imperfectly reciting the alphabet and thinking ourselves possessed of knowledge. In the twelfth century, Hugh of St. Victor knew that "this whole visible world is a book written by the finger of God, that is, created by divine power.... But just as some illiterate man who sees an open book looks at the figures but does not recognize the letters: just so the foolish natural man who does not perceive the things of God outwardly in these visible creatures the appearances but does not inwardly understand the reason. But he who is spiritual and can judge all things, while he considers outwardly the beauty of the work inwardly conceives how marvellous is the wisdom of the Creator." To God's blackboard, then, on which he has scribbled his signature, and study it. Nor need this be mere mystic mumbo jumbo. Do not today's physicists profess to read in the stars mathematical formulae but dimly evident to us, the uninitiated, who see but darkly through our glasses? And don't biologists find in plant and animal the roll of evolution's dice invisible to ordinary mortals?

The mystic hold that inscribed on every created object is its purpose. Fallen and imperfect though the world be, it still remembers Eden and, could we but read the signs, would return us to it bodily if not spiritually. William Coles tells us in his 1656 *Art of Simpling* that, "Though sin and Satan have plunged mankind into an ocean of infirmities, yet the mercy of God, which is over all His workes, maketh grasse to growe upon the mountains and herbes for the use of men, and hath not only stamped upon them a distinct form, but also given them particular signatures, whereby a man may read even in legible characters the use of them." Nature's book will cure your body, so that God's book may cure your soul.

And the prescriptions are writ in language visible to all. Of St. John's wort (*Hypericum*) Coles wrote, "the little holes whereof the leaves of Saint Johns Wort are full, does resemble all the pores of the skin and therefore it is profitable for all hurts and wounds that can happen thereunto. " No longer valued for its alleged dermatological virtues, *Hypericum perforatum* is an import from Europe found in both gardens and waste places—and much touted as an antidepressant. The transparent areas on the leaves that gave the plant its specific name *perforatum* are rich in hypericin, a toxic pigment evolved as a defense against insects. Ironically, several beetles, including *Chrysolina quadrigemina* and *C. hyperici,* now use hypericin to locate and eat *Hypericum.* Which led to their being imported by Australia and the United States to combat out-of-control invasions by *Hypericum*—what they call Klamath or goat weed out West. Out there, *Hypericum* is everywhere. In the tamer East, you're more likely to find it in your neighbor's flower patch. Go on and steal some; it may cheer your soul and let you glimpse, but dimly, the happiness we lost with the apple.

Coles also saw God's hand in the walnut, whose wrinkled halves he thought "the perfect Signatures of the Head," and so prescribed it for headaches. Others think the nuts resemble nuts—testicles—and presume them aphrodisiacs. The Romans threw them at weddings, both to encourage fertility and, it is said, to drown out the noise of the bride's deflowering. Our scientific name, *Juglans,* combines *jovis* and *glans,* the acorn of Jupiter. That *glans* is also the name for what Wikipedia terms "the sensitive bulbous structure at the distal end of the penis" suggests that Jove's acorn may hang elsewhere than from a tree. Michelangelo adorned the Sistine Chapel with nude males around whom he piled an inordinate number of Jove's royal nuts. The pious take these to be coy allusions to his patron, Pope Julius II, born Giuliano della Rovere, Giuliano of the Oak. But since these glans more than resemble the penis glans Michelangelo included on his men, they have come to be known as *teste di cazzi*—penis heads. As we all know, men have two heads, and they think, if ever they think, with the smaller.

Were the silly serpent lurking in trousers or bushes to bite, you'd need a cure. A common American one is rattlesnake plantain (*Goodyera pubescens*), an orchid whose basal rosette of veined leaves is held to resemble the skin of rattlesnakes and hence to cure their bite. Nor was this lunacy confined to European settlers. Indians would "for a trifle allow themselves to be bitten and cure themselves at once," according to one source. Seneca Snakeroot (*Polygala senega*) retains its similar reputation today, one oneline hoodoo site explaining in capital letters lest we grow inattentive, "SENECA SNAKE ROOT, also called MILKWORT and RATTLESNAKE ROOT, protects from SNAKE bite and the work of false friends. It is used in the same way as BUTTON SNAKE ROOT, RATTLESNAKE MASTER, and the like. Worn in the shoes, it protects against all forms of evil messes and tricks. Under the name MILKWORT, it is also said to aid lactation in cows, goats, and women; its taxonomic name, *Polygala*, means "much milk" in Greek. Protective Shoe-Powder: Mix SENECA SNAKE ROOT with any regular brand of foot powder or with Fear Not to Walk Over Evil powder. Dust your feet and inside your SHOES with this, and no tricks laid in the dirt can harm you, for you will walk shielded. We make no claims for SENECA SNAKE ROOT, and sell as a Curio only."

Not all cures were harmless curios. White snakeroot (*Eupatorium rugosum* or *Ageratina altissima*) can kill. Only in the past century did scientists discover that the "milk sickness" that killed scores of settlers in the Ohio Valley—including Abraham Lincoln's mother—resulted from drinking the milk of cows that had eaten snakeroot, which grew in thick clumps in newly cleared pastures. Snakeroot's cousin Virginia snakeroot or serpentaria (*Aristolochia serpentaria*) has a flower that looks rather like a snake rearing its head to bite—reason enough to assume it useful against snakebites. A cousin, the European *Aristolochia rotunda*'s flower, thought to resemble the birth canal, occasioned the common English name "birthwort." And perhaps the Roman name which we still use—*aristolochia, aristos* (best) and *locheia* (childbirth).

Boneset or thoroughwort (*Eupatorium perfoliatum*) and Joe-Pye weed (*Eupatorium purpureum*) were also held efficacious. Boneset's leaves wrap round their stem, leading some to wrap them round broken bones to hasten healing. Fevers such as breakbone—what we call dengue—and yellow plagued the States well into the nineteenth century—and *Eupatorium* was thought to both help break these and heal broken bones. Joe-pye weed—Purple thoroughwort, or gravel root (*Eupatorium purpureum*)—was dragooned to cure kidney stones—or gravel (in which it often grows alongside waterways)—and sore throats, the latter on the basis of its hollow stems.

*Wort* is an Old English word for root or herb, and many were associated with various cures. Louseworts (*Pedicularis*), whose flower heads look—vaguely—like a louse might were he an inch long, repelled lice. *Pedicularis canadensis,* Canadian lousewort, a pretty flower of dry open forests, is our most common lousewort, though the USDA lists thirty-six species found here. Liverworts (*Marchantiophyta*), whose primitive leaf-like structures were thought to resemble livers, naturally cured liver ailments, as did *Hepatica,* whose three-lobed leaves really do recall a liver. You will find *Marchantiophyta* in damp places like seeps and river rocks. Our two varieties of *Hepatica nobilis,* which the naturalist John Burroughs loved: "There are many things left for May, but nothing fairer, if as fair, as the first flower, the hepatica," grow in rich woodlands.

The world is filled with God's mystic messages, and this bouquet of beauties cannot hope to catalogue them all. Let us leave with a plant the wisest of all men, King Solomon, set his seal upon: *Polygonatum,* Solomon's seal, of which we have several species. *Polygonatum,* or many knees, perhaps because the root is knobby kneed. But its common name, Solomon's seal, in Latin, *Sigillum Salomonis,* and French, *sceu de Solomon,* is thought to refer to the round scars on the plant's roots, said to resemble a seal stamped in wax. The Bible tells us that Solomon "spake of trees, from the cedar tree that is in Lebanon even unto the hyssop that springeth out of the wall." And there are those who think the roots' seals, cut properly, reveal in Hebrew his wisdom. *1 Kings* tells us that "king Solomon loved many strange women," that "he had seven hundred wives, princesses, and three hundred concubines," "that his wives turned away his heart after other gods," to Ashtoreth, Milcom, Chemosh, and Molech. Who knows what forgotten wisdom Solomon learned from them and their gods?

But what he taught his wives and princesses and concubines is still, alas, being beaten into women. Observing that Solomon's seal seals broken bones, the English herbalist John Gerard remarked, "the root of Solomons seale stamped, while it is fresh and greene, and applied, taketh away in one night, or two at the most, any bruise, black or blew spots, gotten by fals—or women's wilfulnesse in stumbling upon their hasty husbands fists, or such like." I grow it, of course, in my yard.

# SEX

The discovery that flowers are sex organs titillated the world. Darwin's grandfather wrote an entire book of poems on floral promiscuity. Botanists, then nearly all male, decided, like adolescent boys possessed of their fathers' girlie mags, to organize the world by sexual organ. They still do, which leaves those of us unwilling to practice gynecology upon a daffodil to take their word on such matters.

And learn the names for a perfect flower's perfect parts: the slightly salacious stalk or *peduncle* (my friend Gordon's nickname for his peduncular penis), above which spread the *sepal's* green gown and *petal's* colored petticoats, in whose sweet-scented center rise male and female parts, our gardens the haunt, dear Darwin, of hermaphrodites. The coy female parts, gynoecium ("woman's house"), curve most femininely in the sketches I have peeked at and contain one or, *miracle du Diable,* more than one *carpel,* the holy grail of pollen grain. At the carpel's bottom sits the *ovary,* whose placenta-hiding walls describe as near perfect an arc as did those of her I hiked behind ascending North Carolina's Big Butt Mountain. From this rises a skinny-waisted *style,* the floral equivalent of the vaginal canal. Atop which rests the *stigma,* whose pollen-poaching parts are as various as the flowers they adorn, from the feathered finery of wind-pollinated grasses and trees to the sculpture-garden grotesques visited by various animals. (In Greek, *stigma* means mark or point, and its use for the floral door of delights should not be taken as stigmatizing those who spread petal to wind or bee or bird.) You may know *carpels* as *pistils;* certainly I did before the Internet informed me that I was several decades *dehors courant.* Which is as well, *pistil* coming to us from Latin *pistillum,* pestle, from what excited botanists thought the female organ resembled, unaware as to the proper roles of mortar and pestle in sexual metaphors.

Whirling round most single *pistils*, ahem, *carpels*, attends a crowd of *stamens*, collectively called the *androecium* ("man's house"). Each of these *stamina* (who knew before this, the plural of *stamen*, Latin for "thread") contains a thread-like *filament*, atop which rides the *anther*, where reside pollen, the business end of the *androecium*. Not being a scientist by trade, I note in passing that the Mexican *Lacandonia schismatica* reverses this all-too-familiar-to-human-males arrangement of more males than females. *L. schismatica*'s three or four studly *stamina* standing surrounded by a bevy of, count them, sixty to eighty *pistils* cocked and ready for action.

A small magnifying lens makes much of this world visible. And should voyeur you not spy all the parts, despair not. I did say we were ogling a perfect flower's perfect parts. Some flowers, alas, like us, are less than perfect. Perfect flowers are hermaphrodite, imperfect flowers, like most of us merely male or female, with the other sex's bits and pieces vestigial. In addition, some plants carry flowers of both sexes and are *monoecious* (Greek for "one house"). Others separate the sexes tree by tree and are *dioecious* (Greek for "two houses"). Which explains that berryless holly you paid good money for at Lowe's; it's male and will never bear fruit. Autumn olives and Virginia cedar are also *dioecious;* if you like the one's perfume and the other's Christmas possibilities but deplore their invasive tendencies, merely kill in fall those olives bearing red and those cedars bearing blue fruit (although one female gardener I suggested this to accused me of misogyny). Those perfectly *monoecious* hermaphrodites may sound kinky, but, to Freud's relief, they avoid incest by placing boys in one place, girls in another, or by having one mature earlier than another. Not that Mother Nature can sleep easy at night. Some of her children like to change sex, meaning your Jack-in-the-Pulpit may well be Jill one year and Jack the next. But children are never easy, are they?

Now for the fun part, at least in humans—the mechanics of pollen transfer. Why not let the wind do it, which is what grasses, maples, oaks, pines, and ragweed do? But blow jobs waste seed, so those who willing to so blow must be prodigious in pollen, as anyone who's seen a pond yellow over with pine or oak dust knows. As do hay fever victims who curse ragweed. And for the wind to work, the way must be clear. Grasses enjoy open fields, ragweed waste places; there they pollinate where and whenever the wind blows them. Trees, however, must get their pollen out before leaves appear and block access to all those breeze-dancing stigmata. Which is why hickories, oaks, and walnuts grow tassels in early spring—miniaturized male flowers whose harder-to-spy female counterparts hide where, later, nuts appear. And, done, drop their used and now useless male catkins as if our sidewalks were a lovers' lane.

Less promiscuous are those who choose their lovers, which many a flower, designed to dupe a bug or bird into getting the boy bit into the girl bit, do better than we humans. While what it wears is flower, the magnolia's finery is not as fine as later floral versions. Evolved before bee and butterfly, the magnolia favors beetles as pollinators, a rough-and-tumble crew as likely to bite as kiss you—which explains the magnolia's tough petals. And not all that bright—no need, then, to evolve a fancy flower. Stick to the *Bauplan,* a large, day-blooming, often white or green, bowl-shaped flower with exposed sexual organs, a distinctive odor, and moderate nectar offering. So basic that the petals and sepals are basically the same and so minimally evolved that the bits and pieces are large and easily identifiable to bumbling neophytes such as beetles and myself. Each petal the same, all arranged in a whorl around a conspicuous central stalk with numerous stamens bunched below equally numerous carpels.

Despite their name, tulip poplars are magnolia relatives and their large tulip-looking flowers reminiscent of their cousin. Being more plentiful, to these I go to botanize. The tulip's perfect flowers (with peduncle, Gordon) are erect and cup-shaped, easy to spot lime green and yellow beyond reach high in the forest canopy. Some drop, though, and these fallen few must do, their three green sepals surrounding the six petals which resemble a tulip, whose interior is touched in orange and yellow. As with the magnolia, a crowd of stamina bunch beneath a bevy of carpels crowning a center shaft that matures into a cone of loosely attached woody seeds.

The beetles that pollinate these trees stumble around, defecating in their food, until they accidentally dust themselves with pollen. More evolved flowers designed to attract "smarter" insects (the term's not mine) such as bees have so modified the magnolia's luncheon-plate approach that the flower in violets and iris is now a funnel into which the bee must force herself to obtain nectar, thus transferring pollen. These also provide a landing pad for bees to rest upon. And are often blue—bees' favorite color—and sport nectar guides, floral patterns that, like airport landing lights, say this way to the runway. Orchids have famously taken evolution to an extreme, evolving to be serviced solely by one sort of insect, and some devious enough to themselves resemble insects ready to mate. Our lady's slippers (*Cypripedium* ssp.) have not carried deception that far, but their come-hither color and sweet scent deceive bees into thinking there's something even sweeter awaiting them inside the slipper. Alas, no nectar waits, and, disappointed, they stumble out, dusted with pollen. But the busy bee is no bumbling beetle. Many learn to ignore the hussy orchid, and her numbers are few.

I once knew a man who counseled wearing Old Spice aftershave on the assumption it would remind women of their fathers and deceive them into thinking I had their best interests at heart. My luck was to wind up dating one who hated her father and all men doused in his aftershave. Which is by way of introducing flowers that we think stink but to which flies fly. Such as the sex-changing Jack-in-the-Pulpit (*Arisaema triphyllum*), whose pulpit is a modified leaf—spathe—and Jack (or is it Jill) a spadix bearing numerous small flowers. The plant's mottled stems are thought to resemble rotting flesh, which, coupled with a mushroomy odor, attracts carrion flies that find their way in and, disappointed, leave to carry pollen to another gay deceiver. Jack so builds his spathe that there is a small opening at its base through which flies flee. But Jill omits the hole, and many a fly dies forlorn in her nether lands. An even more odiferous cousin has earned the moniker "skunk cabbage" (*Symplocarpus foetidus*) because it smells so bad to us. But insects enamored of roadkill and putrescence flock to it in late winter, when its self-generated heat melts the ice and snow around it. Pawpaws (*Asimina triloba*) also depend on flies, their small, earthward-facing maroon flowers developing a, to us, rank odor. These begin as females, turning male as they age, open, and—but of course—smell worse. You can, like a fly, tell which is which by whether the demure female stigma or pushy, numerous male anthers are visible—something pawpaw lovers learn, since the plants are notoriously finicky fruit setters.

Hummingbirds famously love red, a color bees seem indifferent to. And so we are counseled to plant red blooming flowers, and our feeders are red, red, red. Flowers adapted to the long-billed ruby-throated hummingbird's tastes are long and tubular and red or orange. The bird inserts its bill into a flower, flicking its tongue in and out to lap up nectar, while pollen it will transfer to the next flower falls on it. These are, in botanyspeak, ornithophilous (bird-loving) flowers such as the spring-blooming spires of red buckeye (*Aesculus pavia*) and the nodding spurs of our red and yellow columbine (*Aquilegia canadensis*). Hummingbirds stoking up for fall migration may play lascivious with red trumpet honeysuckle (*Lonicera sempervirens*), various bee balms (*Monarda*), the bright cardinal flower (*Lobelia cardinalis*), orange trumpet creeper (*Campsis radicans*), and orange jewel weed (*Impatiens capensis*).

While caterpillars may depend upon one or two plant species for food, butterflies are less choosy, sipping nectar from any flower colorful and fragrant enough to attract them and with broad flat surfaces to perch on. The monarch caterpillar, for example, dines as exclusively on milkweed as my son did on Spaghetti-Os. Grown, both monarch and Edward developed more catholic tastes. Adult monarch butterflies sip from natives asters, butterfly, iron, and milkweeds, cone flowers, goldenrod, and thistles. And, dung, seeking salts and nutrients not found in flowers. Happily, a trait my son does not share.

As butterflies flit off to bed, moths replace them, seeking fragrant, tubular flowers filled with nectar, such as the Jimson weed, evening primrose, and *Nicotinia* of my backyard. The widest spread of our hawk moths is the five-spotted (*Manduca quinquemaculata*), whose caterpillar is the infamous tomato hornworm. But whose adult form flying from flower to flower in the growing dark almost (almost, I said) pardons the sins of the child. With tongues up to fourteen inches long, hawk moths are supremely adapted to feeding on long-spurred flowers. When botanists puzzled over what could possibly pollinate the foot-long nectar spurs of a Madagascar star orchid, Darwin suggested a hawk moth. Pooh-poohed while alive, he proved to be right. Even more intimate is the love affair between the yucca and its moth. *Yucca filamentosa* depends entirely upon *Tegeticula yuccasella* for survival. After laying eggs in the yucca's ovary, the moth packs it with pollen, thus fertilizing the ovary, whose seeds her larvae will eat. Should they eat too many, the yucca aborts both fruit and larvae. Less voracious larvae live to reproduce, moth and yucca playing boudoir politics millennium after millennium.

And it seems millennia since I too sported in the shade with amaryllis. But there she is, and I, enamored of her scent, her color, shape divine, must, like moth, or carrion or butter fly, bee, bird, or breeze, leave now and scatter pollen if she will her *gynoecium* open to my knock knocking at her door.

# BREEZES

Forest shadblow and box stores' six-packs of greenhoused tomatoes already in flower fool some into thinking summer's here, especially after a long, cold winter that dragged February into April. But spring chills linger long, and plants native to southern Mexico should sweaters wear when going out May nights in the Shenandoah. As many a transplant here has learned standing amid the crumpled ruins of their garden following a hard frost. Leastwise in the Valley. Yet from his elevated perch aside House Mountain, Philip Clayton spies summer coming two weeks before the Valley-bound, and he plants tomatoes when we're still hunting morels under oak leaves the size of mouse ears.

Lift up your eyes unto the hills and see spring climbing the Valley's flanks. Though we may be greening out down here, the mountains' sides are barely yellow, and the trees atop still bear the gray of winter. Seekers of trillium and Solomon seal prolong their search by climbing upslope, spring full blown here but buds but barely blooming two hundred feet higher. Should they carry a thermometer with them, they'd find the temperatures lower as they rise, five degrees for every thousand feet, a blessing in hot August, a curse in frozen February.

So why, if five degrees colder than the Valley, do Philip Clayton's tomatoes prosper when ours freeze? Because of the Valley breeze, which you'll notice if, lulled by a warm day, you walk the fields at night without a jacket. Then you feel against your bare neck a breeze as chill as that when someone steps upon your grave, but steadier, long into the night, enough to send you seeking hay bales to hide behind. Such downslope, katabic breezes arise when the night air over the mountains, being nearer ground, loses heat more rapidly than adjacent air at the same altitude over the Valley. Colder and heavier, the air flows as a downhill breeze. The physics is similar to that which caused the sea breezes of my youth, which still blow through my memory of languishing in sweltering heat on porches pitched to catch a breeze that blew away heat and mosquitoes. These were but breezes. Katabatic winds off the ice sheets of Greenland and Antarctica, called Piteraq and williwaw, can build to hurricanes 140 mph.

The same physics reverses flow, so we might fly in gliders five hundred miles along the Appalachians' flanks. An anabatic, upslope wind arises on calm, sunny days when a hill or mountaintop warms more than air at the same altitude over a valley. A breeze rises as the valley air flows upslope, the warmer air temperature causing a lower air pressure that draws in air. Clouds forming over the mountains as air rises into cooler regions can betray the presence of such breezes, and many a day the Valley will bask in sun while clouds rise over higher ridges.

Downslope, katabatic physics can turn the Valley into a vast frost pocket, some places deeper pocketed than others. Although a thousand feet higher than Lexington, Philip enjoys a growing season weeks longer than my city one, and I enjoy one longer than does Lacie, around whose log cabin Davis Hill's chill breath can still kill a week after I've given up watching in fear the outdoor thermometer. The spring she drinks from has cut a bowl in which her garden sits and where collect the killing frosts of May and September. Microclimes, the master gardeners call these, which you must learn if planting tender vegetables.

Greek *katabasis*—downgoing—and *anabasis*—upgoing—gave us these breeze's names. But the word's most famous use is Xenephon's *Anabasis,* his account of the march of the Ten Thousand, an army of Greek mercenaries, into and out of the Persian Empire, best known for the moment when, upon seeing the Black Sea, they cried, "*thálatta, thálatta,*" the sea, the sea. The Shenandoah's Sea dried long ago, and even now she receives less water than the rest of Virginia, thanks in large part to the same physics that blasts Valley tomatoes. The great torrents of air flowing west to east across America must rise to cross California's Coastal Range, the Sierras, the Rockies, and the Appalachians. Each time they rise to higher elevations, their waters condense and fall, in the fogs and rains that feed the redwoods, the snows that Los Angeles drinks, and those the rich ski at Vail and Breckenridge. The Appalachians too arrest the eastward flow of moisture laden air, which falls in record inches in the Smokies and along the western edges of the Cumberland

Plateau. By the time the winds reach the Shenandoah, they have dried, and Lexington's thirty-eight annual inches of rain are four less than Virginia's average. The Blue Ridge do their best as well to squeeze moisture out of fronts approaching from the east, and Lynchburg, only thirty-five miles east of Lexington, gets five inches more rain than we do.

Such prevailing wind and rain patterns govern weather everywhere. The Caribbean's Leeward and Windward islands take their names from their relations to the trade winds, and places in Hispaniola's windward Dominican Republic drown in more than one hundred inches of rain while those in leeward Haiti parch with sixteen. The Himalayas that drench Indian villages in 467 inches of rain permit only 4.3 to fall on inner Tibet. Such shadows may not rule an individual mountainside in the Shenandoah, but the lay of the land affects what grows upon a mountain. To understand fully the sun's parching possibilities, it's best to walk Lexington's sidewalks in July's stultifying heat. You'll seek shadows and plot courses that avoid south- and west-facing streets. So too do plants. Although it's unwise to trust that moss grows thickest on a tree's north side, you can sometimes discern a prejudice for moisture. That Philip Clayton's slopes face south helps explain both his frostfree growing season and his forest, dominated by oaks, hickories, and pines. These trees endure drier lands than those on House Mountain's shadier, species-richer north-facing slopes, where maples, beeches, tulips, basswood, oak, and magnolia hung with *Aristolochia* vines can reach sizes large enough to let you dream you're in an ancient wood.

Spring wildflowers adore such deep soiled, well-watered slopes. Two miles downslope from Philip's dry, blue- and blackberried understory, the Balls gaze toward a north-facing slope white with trillium in spring. On the Appalachian Trail over Thunder Ridge you can hike for ten miles through such trillium, may apple, Solomon's seal, bellwort, and a dozen kinds of violet sheltered from the south and west's desiccating sun and wind. But turn a ridge, and the forest floor bares, tulips gone and oak took over. Turn yet again another ridge, you're back in flower fairy land. And so it goes in May, the land's alternating lays amusing, bemusing, as, climbing, you leave spring below, the weeks rewind, and you find March and winter lingering where the mountain faces north, the leaves in buds still hugging themselves for warmth, the-first-to-bloom-in-spring bloodroot barely daring to poke a leaf-wrapped stem above last year's litter, and not a trillium in sight. Come August, when even the hickories thirst for water in the heat, everything living flees the sun, and Philip Clayton, watering tomatoes, wishes perhaps he lived on Davis Hill, where frosts come early and leave late. On such dog days, it's best to imitate the Greeks and go anabasis in the mountains, where leaves stir and humans glide on breezes those in valleys cannot feel.

# PINING FOR THE PAST

Edward has placed on top of the hall wardrobe the stubby skeleton of a pine tree, all knob and swell and not a straight inch from which to cut board. Heartwood, enduring as parents' memories of their offspring. Splinters of such resin-soaked wood burn like matches, and as kids we gathered and split their skeletons as Christmas presents for the pyromaniacs among us. When the workmen cut my floors to install heat, they sawed through inch-thick boards of such heart pine, a slab of which I've saved to inhale like church incense on my worst days, there no stronger tie to my childhood than the odor of a pine's heart broke and bleeding resin. Cedar's a close second, redolent of sweaters pulled from chests and Christmas trees. But it's light-wood, the heavy heart of pine, that takes me back to the too-large campfires of my youth. My campfires now are paltry things, the few twigs I need to boil a can of water. But even they, reduced, smell ever so faintly of lost youth and forests.

Hardwood trunks litter forest floors, hollow-hearted oaks whose wooden skins reverberate like drums when beaten. Though termed softwoods, pines build their hearts of sterner stuff than oak, and, outer husks of bark and trunk rotted back to earth, their grainy hearts remain half-buried in leaf litter, refusing, like lost love, to leave no trace. Shrunk and wizened essence of pine, its platonic form made manifest, fresh split lightwood reveals a grain tinted rich orange red by the congealed blood of *Piniceae*. It was slivers of these the Indians of eastern North Carolina stuck in John Lawson's skin and lit him like a birthday cake to burn alive. To this day, the wood smokes black and greasy and can so coat a flue with residue the chimney burns.

Fungal heartrot can infect the longleaves of Carolina, and these hollow-hearted the red cockaded woodpecker seeks out by some olfactory magic—the only woodpecker in the world to make its home in living wood. Whose resin, weeping from the wound the bird has made, bleeds downtrunk in half-congealed long lines of grief, thought to discourage foraging snakes that dream of baby birds for breakfast. Virginia's pines resist rot, and those dead trunks that stand are targets of pileated woodpeckers, now the world's largest since we killed off the ivory billed. You can hear them in the deep woods, crying like crazed chickens, their rat-a-tat-tat-tatting ricocheting through the forest, and catch their uneven flight as they sine wave through the half-gloom of mature trees, a flash of red head blood bright atop a body robed in priestly black and white. They leave in their wake inch-long chips of white wood scattered at the base of trunks they scavenge for bug and beetle. Take home a pocketful, if only to look at when you think to pound your head against the walls closing in and begin yourself to cry like a half-crazed pullet.

Calmer are the memories cedar heartwood evokes, ghosts of Christmas past and Christmas trees long stripped of ornaments and burned on Twelfth Night, wool sweaters fresh from summer storage, and the broken twigs from back when you thought you might escape whatever haunted you by climbing up a tree. Now come upon one grown grand enough to have killed off its shade-choked lower branches and break a thin-as-a-grade-school-pencil twig and sniff the maroony red where harbor memories you seek to plant in those you love by storing their sweaters in chests made of such wood and taking them to farmers' fields to find a cedar Christmas tree.

These have needles fashioned to survive several years, unlike the profligate deciduous that throw theirs off come summer's end. Yet evergreens too lose their leaves. You can witness leaf fall easiest in white pines, whose older needles brown up and fall as if the tree were sick. Pines bundle needles in packages of two or three or five according to species, a quick and dirty guide to identification. Pick apart a bundle—fascicle—and regard a leaf whose reduced surface area is designed to withstand the lacerating winds of winter, whose waxy cuticle prevents the overloss of water in winter, when frozen ground makes water scarce. With less surface area, a needle may convert per minute less sun to food than broader deciduous rivals, but, like childhood's Little Engine That Could, its steady chugging outdoes the flashier competition. Which in part explains the triumph of conifers in Canada and Russia's boreal expanses.

The boreal gasps its last etiolated breath in the spruce- and fir-tipped summits of the Appalachians. A thicket of young trees is nigh impenetrable, thanks to pendulous branches designed to shed snow and support each other one upon the other. A lesson Edward and I learned when we trimmed the hemlock out back so that we might sit beneath its summer shade. We forgot that, come winter's wet snow, the upper branches, unsupported, would bend and break and leave us shadeless in July. Such is the origin of our kindergarten visions of conifers, each a perfect triangle of supporting branches. Like rambunctious kindergarteners, white pine will rebel against rules designed for its own good and rival branches fight for the lead. Then trees grow like candelabra, begging wind and ice to bend, break, and blow out their green candles, so that, come spring, the forest floor is littered with the sticky stumps of failure.

And cones, the enduring husks that package the seeds of conifers and beg collecting. Edward once had such a museum of the conifer, with cones collected from across the country, the redwood's unprepossessing miniature beside the sugar cone's impressive eighteen inches, the hemlock's light touch next to the longleaves' weighty footlong, the fir's fragile upright falling to pieces beside the spruce's sturdier tight-scaled cylinder and the white pine's sticky look alike, the Virginia pine's stubby, prickled fortress that, like its Table Mountain cousin, refuses to release its hold upon a branch. And all, if still limber, opening and closing with the humidity, twenty sundry weather stations forecasting sun or rain. We still have hung beside the door a greenish loblolly cone from Carolina so that we need not consult the radio before venturing out. Such moisture-moderated movement seems designed to permit seeds to fall when most advantageous. Like us, pines take pains to see their offspring well off, and many grow their cones high up and on the branches' ends so seed might twirl far off upon a breeze.

Thick walled and prickled, pine cones testify, like sea shells, to a hostile world where bird and squirrel and chipmunk attack. Unlike angiosperms, which dress their seeds in flesh designed to be eaten, conifers guard their naked seed—gymnosperm in botanical Latin, *gymno* meaning naked, an etymology the seeds share with gymnasium, the room where naked Greeks exercised. The cones you find upon the ground have few such seeds; these have already fled and what you hold the empty husk of future forests. Some cones, like those that will not release their grip upon the Table Mountain pine, open only after fires, and seeds fill these, if you've the patience to tear them cone from limb and scale from cone. Buck-toothed squirrels are better huskers than I, and many a day I've come upon their picnic litter in a pine grove, the carcass of a cone browning in the sun, surrounded by bits and pieces of half-gnawed wood I hope more satisfactory than the pencils I gnawed through grade-school afternoons.

From town, I can lift up mine eyes and see overcoats of evergreen draping the rocky ridges of surrounding mountains. Should you climb these bony backbones of Virginia, you'll find her namesake pines, scrubby, scrawny, branchy trees so unfit for timbering they're left alone for folk like me and squirrels to wander through. They grow happily on these thin-soiled and bone-dry ridges, useless for anything more serious than wasting July away and tasting tart upon the tongue the green dream of a needle.

# BURIAL ALIVE

Sunning on a rock beside busy Brown Mountain Creek happily burbling about having cut a channel in its stony mattress, I lazily noticed the rock's undercut underside fretted with six-inch tubes of mud. Nurseries of the organ pipe mud dauber (*Trypoxylon politum*), whose female constructs her panpipe nursery from mud balls that she carries to a likely spot and then mudpies into empty cylinders awaiting filling. Tedious, yes, but mother love is something to admire, and you can tell from the sometimes multicolored muds that make up a panpipe the degree of maternal devotion that went into constructing this tomb for the living. Tomb because, completed, these pipes hum not with baby babble but with the subdued SOSes of paralyzed orb-weaving spiders buried in mud tunnels of slow death, where they are eaten alive by the so-slow devouring jaws of juvenile mud daubers feeding on still-conscious spiders. Sated, these vampire larvae spin themselves a silk cocoon, mature, and dig to freedom and to light, leaving in their basinets the legs of those they ate.

Kin to these is the black and yellow mud dauber (*Sceliphron caementarium*), whose chunkier two-inch long nurseries decorate the back porches of America. Macabre mothers of malfeasance, these black and yellows and organ-piped perverters of God's favored instrument are visited—God being, though distant, just—by the iridescent blue mud dauber (*Halybion californicum*), who, despite its Latin name, you'll also find on your back porch in Virginia. And who, conscience untroubled, will use a previously constructed organ pipe as nursery, waiting until it's vacant. Or, need greater than morality, excavate and damn to die the previous inhabitants in favor of her own offspring and their delights, the delectable black widow spider. Pause, then, before you beat off with a broom stick the mud dauber's muddy manger; it may be ridding you of widows bitter with poison.

Upon feeling a trapped insect struggling in their web, black widows (we've three of the thirty-some *Latrodectus* species found worldwide) rush out and immobilize the animal by wrapping it in silk. They then poison their prey by biting it. Death can take a decade of minutes, during which time the hapless bug no doubt regards with regret the red sands running out within the widow's abdominal hourglass. Her prey quit moving, the merry widow squirts digestive fluids into its wound and laps her dinner up. Many a spider species wraps its prey in silken shrouds, which hang like warnings of our own impending ends when we walk life's dark woods. Despite Isaiah's promises that webs shall not become our garments, we know, like Edgar Allan Poe, what silken shroud is spun for those buried alive: "The unendurable oppression of the lungs—the stifling fumes . . . —the clinging to the death garments—the rigid embrace of the narrow house . . . the silence like a sea that overwhelms— the unseen but palpable presence of the Conqueror Worm— . . . We know of nothing so agonizing upon Earth—we can dream of nothing half so hideous in the realms of the nethermost Hell."

Rome proclaimed both that no one might be buried alive in the city and that a Vestal Virgin proved unchaste must there buried be alive. Solomon-like, the city resolved the contradiction by dressing the offending female like a corpse, reciting over her the burial of the dead, and entombing her alive in a vault containing food and water sufficient to avoid the law. Later despotisms condemned as well the adulterous, the female murderer, and the thief to slow asphyxiation. And the Japanese in Nanking, the Nazis in Europe, Mao in China, the Khmer Rouge in Cambodia, all practiced Poe's premeditated murder, burial alive, and no Lazarus rose from the grave to give the lie to oppression.

But the dead do live and, living, rise, as broods of the *Magicicada* periodical cicada testify every seventeen years when, arising from sullen earth, they hymn to heaven the mating song I heard hummed beneath the brighter measures of Brown Mountain's creekbed lyrics. Even lower than the low-toned tenors of *Magicicada* is the buzzing of *Sphecius speciosus*, the cicada killer, a wasp large enough to frighten Paul Bunyan. And one that preys exclusively on cicada, carrying them one by one to her subterranean burrow, where larvae feed on the paralyzed. Cicada killers are, in turn, parasitized by velvet ants, a burrowing wasp (*Dasymutilla occidentalis*), whose pupating larvae feast upon cicada killers feasting on cicada martyrs singing as they die.

But we need not wait until our body dies to feel the brush of shroud upon our face. Who has not hung asphyxiating in the web of work wishing for a savior to rid him of the thing that lurks within the corner office? At his friend Karl Marx's burial in London's Highgate Cemetery, Frederick Engels cried, "Marx was the best hated and most calumniated man of his time. Governments, both absolutist and republican, deported him from their territories. Bourgeois, whether conservative or ultra-democratic, vied with one another in heaping slanders upon him. All this he brushed aside as though it were a cobweb, ignoring it, answering only when extreme necessity compelled him. And he died beloved, revered and mourned by millions of revolutionary fellow workers—from the mines of Siberia to California, in all parts of Europe and America—and I make bold to say that, though he may have had many opponents, he had hardly one personal enemy. His name will endure through the ages, and so also will his work." Look on his works, ye mighty, and despair—spiders now web the unread *Manifesto*, buried under the piled-high slanders of bourgeois such as you and me. Atheist Engels presumably had no use for Bildad's cold comforting of Job: "How long wilt thou speak these things? and how long shall the words of thy mouth be like a strong wind? . . . Whose hope shall be cut off, and whose trust shall be a spider's web."

With friends like these, smote with boils he scrapes with a broken pot, Jobs turn to the hope of marriage. And hear a harridan urging, "curse God, and die." As I heard a woman once, deliberately misquoting Emily Dickinson, prophesy to all assembled that marriage is "all we know of heaven, and all we need of hell." Hell it must have been—she parting ways with him to whom she'd sworn "whither thou goest, I will go; and where thou lodgest, I will lodge: thy people shall be my people, and thy God my God: Where thou diest, will I die, and there will I be buried." Her former husband lies where all lost loves lie, in her heart, wander where she will. Like the wasps that plague our summer afternoons are buzzing memories we cannot ignore, the origin of the complaint we men sing to ourselves: who's kissing her now?

Who's kissing her now? Death, my friend, death kisses her, as he will you. Do you not, even now, feel his breath upon your neck's nape? Shiver with cold despite August's heat? Would you, wrapped in thoughts of death, be resurrected, come forth like Lazarus to Christ's tears? Lazarus, they say, remembering his four days in hell surrounded by the souls of those who would not rise, never smiled again despite a long life lived preaching the resurrection of the body. He knew, like Sylvia Plath, what horrors await those who would reanimate the dead:

Herr God, Herr Lucifer
Beware
Beware.

Out of the ash
I rise with my red hair
And I eat men like air.

# SHELVING

Many a tree has found its way onto a shelf by turning into a book. And many a shelf has found its way into a tree. These are the shelf fungi, a group not catalogued by DNA or Dewey Decimal or Library of Congress but linked only by look and habitat—the trunks of dead and dying trees. And, like the books I used to sit me on the floor of library stacks to read, these fungi can reveal much about the world we walk.

Aging black locusts often sport woody shelves nigh near impossible to break off. These are *Phellinus robiniae*, parasites of the trees' heartwood. Though nearly all locust fence posts grow hollow from the white rot such fungi cause, the posts' outer layers endure and harden, an old locust post something that daunts even chain saws. *Phellinus* means "corky," referring to the fruiting caps, which crack and furrow with age. Almost as enduring as the black locust they grow on (*Robinia pseudoacacia*, hence the fungus' species epithet, *robineae*), darkened with age, they clothe themselves with moss and lichen. And you'll need a knife or saw to break cleanly off one of these.

More easily obtained, albeit less common, is hemlock varnish, a shelf fungus that makes beautiful in death the hemlocks along Brown Mountain Creek. Although *Ganoderma tsuega* eats the dead, it also cures the living. In China, they boil it and its close cousin *G. lucidum* to make a bitter but reputed beneficial tonic. And in Japan they grow it for its anticancer properties. So sovereign a remedy is what the Chinese call *lingzhi* that you can buy five hundred dried grams (17.5 ounces) online for the reduced price of $150. Or wander the woods and find your own. Its beauty ravished even the mycologists who named it; *Ganoderma* means "shining skin," and *lucidum* also means "shining." Collect one to display upon your mantle until it dries and withers, then boil and drink what's left.

More edible is the oyster mushroom (*Pleurotus ostreatus*), which—vaguely
—resembles a large oyster. Once known, this mild-flavored and firm-fleshed
mushroom is easily recognized, and Lacie, who learned mushrooming from
her German parents, was delighted to find banks of oysters breaking bark on
the stump of a large paper mulberry in her yard. For several years now, she has
gone oystering, a cold and muddy deed done in the dead of winter on the sea-
coast, but an easy snicker snack of vorpal blade in the Shenandoah when they
appear after rains. While we sauté and eat the oyster's fruiting body, the fungus itself, a mat of threadlike mycelia that digest dead wood for its nutrients,
eats nematodes. The mycelia exude a toxin that stuns the microscopic worms,
after which the fungus crawls in through the animal's mouth and anus and
devours the still-living creature—presumably for nitrogen. You can buy jars
of dried *Pleurotus,* or logs inoculated with their mycelium. But cheaper and
more fun is to find your own in the woods and take up farming them on straw
or rotting wood.

Since mycologists warn people like me against picking poisonous look-
alikes to oyster mushrooms, a less deadly choice might be the chicken-of-
the-woods (*Laetiporus sulphureus*). Also called sulphur shelf after its bright
yellow to orange color and almost—almost—impossible to confuse with another mushroom, chickens-of-the-woods can produce more than one hundred pounds of edible mushroom per tree. So bright that they seem to shine
through the woods, they grow shelf upon stacked shelf up hardwoods, whose
heartwood they destroy. Lacie and I have come back from hikes with two
backpacks and four grocery bags full, having had to leave as much or more
behind us for another day. Said to taste like, what else, chicken, *Laetiporus* is
best eaten young and still rubberly tender; older pieces, too tough to eat, can
be added as flavoring to stews and soups.

Should you be more of a red-meat sort of diner, you might prefer beefsteak
or ox tongue (*Fistulina hepatica*). Growing both here and in Europe, where it
was—and is—eaten as a meat substitute, beefsteak fungus really does look
like a slab of meat or liver (hence the species name, *hepatica*), even bleeding
a dull red juice when cut. Feeding primarily on oaks, *Fistulina* causes brown
rot and is said to be the hollower of many an oak. My friend Alan, a long-
time vegetarian, can feast himself on thick slabs of beefsteak, slow cooked to
tenderize a sometimes tough cut of fungus, and think himself more virtuous

than me—although even the best-raised of *Fistulina* taste nothing like a real steak, *saignant.*

One of my favorite shelf fungi—if only because of ubiquity and ease of identification—is turkey tail (*Trametes versicolor*), which causes white rot in hardwoods. If you've hiked the woods, you've seen turkey tail or some look-alike imitator. Fear not: no one eats these thin-as-a-checkbook fungi; we merely admire their multicolored beauty, betrayed in white, brown, yellow, bluish, and cinnamon intershelving shelves occasionally greened with algae. Useless in these United States for other than mantle cluttering, turkey tails are sovereign remedies for cancer in China.

Horse hoof fungus (*Fomes fomentarius*) really does look like a whitish horse hoof someone cut off and stuck to a tree. What's most fascinating about this inedible fungus is its long use. Soaked, cut into strips, and beaten, the dried and treated spongy inner layer, called amadou, has been made into curiosities such as picture frames and hats, examples of which you can view online. *Fomes fomentarius* means "tinder used as tinder," saluting its long use in starting fires: Otzi the Iceman was carrying amadou tinder when murdered five thousand years ago in the Alps.

Though long a favorite of lovelorn such as myself, the artist's shelf (*Ganoderma applanatum*) has fallen into ill repute since having been found complicit in the death of Anne Frank's horse chestnut tree, of which she wrote, "As long as this exists, I thought, and I may live to see it, this sunshine, the cloudless skies, while this lasts I cannot be unhappy." Anne died in March 1945, her tree on August 23, 2010. An eater of dead—and sometimes living—wood, *Ganoderma* forms woody brackets enduring decades and solid enough to climb on. The sporeside underneath of *Ganoderma* provides would-be artists a white canvas to scratch on. Its upper side, like most shelf fungi, betrays its age in easily counted growth rings, the youngest being the outermost and closest to the ground. Most shelf fungi exhibit positive gravitropism, which means they grow earthward. And often betray their growth in yearly—sometimes twice yearly—growth rings. Should the tree they're eating to death have, like Anne Frank's, the bad taste to fall, this gravitropism will cause the fallen shelves to reorient, turn their pores earthward, and continue eating. *Ganoderma*'s nothing if not prolific; one fungus site claims a single shelf will spew 1.25 billion spores an hour—for half a year.

The off-white to brownish dryad's saddle (*Polyporus squamosus*, or many pored and scaly) shelf fungus grows on stumps and logs in the fall. Young shelves can be sliced and sautéd, but older ones are tough and often maggoty —much like people. More fun might be to make paper if you've enough brackets. Recipes online call for blenderizing the saddle with enough water to make a marshmellowy paste you then spread on a flat sieve to dry.

The dryads whose saddles you thus destroy are tree nymphs. Originally associated with oaks, they now inhabit any species of tree and will, it is said, attack with axes those who threaten their habitation. The Meliai were ash tree nymphs, so named after the sweetish sap that oozes from the European *Fraxinus ornus*, it and bee honey (*meli*) thought in pagan times to be ambrosia of the gods. Later, Christians would think the sap of weeping trees a form of heavenly manna, conflating the ash's sap and honeydew, the liquid excrement of tree aphids. You can still buy ash-tree manna should you be in Sicily. Or lick for free diarrhetic manna from forest foliage.

Especially sacred were the *Hamadryádes*, nymphs so bound to the tree they lived in that they died with their tree. There were eight kinds, associated with dogwoods, elms, figs, grapes, mulberries, oaks, poplars, and walnuts. Well on our way to driving to extinction the hamadryades of dogwood and elm, we should walk wary of axe-wielding viragos in our woods. Religion was, for the Roman Cicero, *cultus deorum*, the cultivation of the gods. A sentiment St. Augustine echoed when he asserted that "*religio* is nothing other than the *cultus* of God." We uncultured barbarians forget to cultivate the gods and goddesses who inhabit the living houses of the forest, trees we too often see as so many board feet, whose health, like ours, we compromise with irreligious living and seek to cure by eating the fungi that flourish on the trees we fell. Once the habitat of living gods, our forests die, the dryads disappear, their saddles empty and gone to maggot and rot. Like our souls.

# INDICATOR SPECIES

Run your tongue long round the phrase "boreal forest" and you'll start feeling frostbite in your extremities, be possessed of the urge to reread Jack London's Yukon tales, and take to looking for Christmas trees in June. Boreal descends to us from Boreas, Greek god of the north wind, who comes seeking in December his misplaced offspring, the spruce-fir forests of the highest summits of the southern Appalachians. These rise above five thousand feet to where the South thins in air last exhaled sometime in an Ice Age and sky as dark blue as a Yankee uniform looks down upon the gray fogs perpetually muffling these mountains.

Such postage stamp-sized relics of the Pleistocene, in retreat for eighteen thousand years, are doomed. The state park atop North Carolina's Mount Mitchell, the East Coast's highest point, topping out at 6,684 feet, measures fewer than two thousand acres. All told, there are but one hundred square miles of this southern version of Russia's endless taiga. And it's dying even as you read. From Mitchell's watch tower, one looks down upon a forest of the dead, a hundred thousand thousand endemic firs killed by the balsam wooly adelgid, an accidental European import, by air pollution blown in from the Midwest bathing the summit in water acid as vinegar, and, year by year, by a temperature that inches higher. Warmer now though it be, January still averages but 25° Fahrenheit atop Mount Mitchell, where, in 1985, it fell to -34. Heavy snows pile up December through March, flurries fall in August, and winds 178 mph howl at those foolhardy enough to come in the wrong season.

The dead are Fraser firs (*Abies fraseri*), named for the Scottish plant collector John Fraser, who tramped these mountains two hundred years ago. Called "she" trees by settlers who could milk them for their resin, the Fraser has paired herself with the "he" tree red spruce (*Picea rubens*), which you'll find all the way north to Quebec. But Frasers, though close cousins to the wider spread balsam fir (*Abies balsamea*), are purely southern, growing naturally only as far north as Mount Rogers in southwest Virginia. Odds are that that Christmas tree whose odor so takes you is a Fraser. I scout sidewalks the week after Christmas for discarded Frasers, harvesting their needles into a sock I then hide in my bureau drawer so I might climb the Southern Appalachians every time I change my underwear.

The Fraser fir is what they call an indicator species, a plant or animal so typical of a particular ecological community that, finding it, you can with some assurance say, yes, I am standing in the Southern Appalachian spruce-fir forest. Or on a Christmas tree farm in Pennsylvania, Quebec, or Scotland, where they're grown like vegetables. To find Fraser's companion, the red spruce, I pilgrimage to Spruce Knob, at 4,863 feet West Virginia's tallest peak, its upper slopes shrouded in fog and spruce, an Ice Age relic you can, like North Carolina's Mount Mitchell, drive to. Perched on the steeply dropping Allegheny Front, the Knob's spruce have been so bonsaied by prevailing winds they've lost their windward branches and seem to be green pennants flapping leeward in the wind. Flag trees in English, gnarled and crawling for shelter, in German they're called *Krummholz,* "crooked wood," and are a sign of adverse conditions and a nearby tree line. Though higher at 5,729 feet, Virginia's Mount Rogers's spruce-fir crowned summit is less tortured, growing green and thick and so tall you see nothing but tree and would never know— except for Fraser's fragrance—you're nearer God than any place else in the Commonwealth.

The mountains roundabout Lexington are too worn down to harbor Fraser firs, but the 3,372-foot-high crown of Bluff Mountain, which you can just spy from town, is green with spruce. These, though, are Norwegian fir transplants, happy enough to be reproducing down the Appalachian Trail to behind Punchbowl Shelter, where salamanders share a pond with hikers. They shade a marker to four-year-old Ottie Cline Powell, who died here, seven miles from the school he wandered away from in 1891. Though not the legendary spruce-fir, logged out and harboring a grove of nonnatives, Bluff Mountain is still more northern than southern. Or so the experts say, who call what grows along the Appalachians' southern spine an extension of the northern hardwood forest. Which, it turns out, might be dominated by hemlocks and white pines, or basswood and buckeye, or oak and hickory, depending on how much moisture there is, what's underground, and which way the slope faces.

Locals, however, know they're in a high up forest when they spy striped maple (*Acer pensylvanicum*), called so for its green and white striped bark and also known as moosewood farther north where there are—or were—moose. *A. pensylvanicum* begins at around 2,000 feet on many nearby mountains, and, indicator species that it is, tells you you're a welcome halfway to the top of, say, House or Bluff or Jump or North Mountain when you first spy its stripes in the understory gloom. Never very large, the maple makes as pretty a lightweight walking stick as you might like, either with its striped bark or stripped bare and white. It's native to woods from southern Canada south to Ohio and New Jersey, reaching into Dixie along the narrow Appalachian spine. The only one of its kind in North America, it looks to Asia for relatives, where some twenty "snakebark" cousins grow from the Himalayas to Japan. The three-lobed leaves, often quite large, exhibit much the same structural characteristics as I discuss in "Maple Leaf Drive," while its pale green flowers hang in a drooping raceme that transforms by fall into a chain of paired and winged seeds—samaras—which, over winter, helicopter away to start new trees. Although I visit striped maples too infrequently to have noticed, the U.S. Department of Agriculture's experts say individual trees can change their gender, 27 of 243 sampled swapping roles, mostly male to female.

Male or female, you'll find the richest forests in North America in Appalachian coves, where 200 million years of uninterrupted evolution have spawned and protected species whose numbers rival those found in rainforests. Here grow ash, basswood, beech, black cherry, buckeye, cucumber and fraser and umbrella magnolia, hemlock, hickory, white pine, silverbell, sugar and red and striped maple, tulip poplar, white and red oak, and yellow and black birch, the world's richest array of salamanders and mussels, and birds flown in from South America to nest—the scarlet tanager, dyed deeper red than cardinals of Valley or Vatican; the nearly vanished cerulean warbler, like a bit of blue sky dropped into the trees; the common black and white warbler, striped as if a bit of gneiss had taken wing and flew. Where a cove steepens, the soil thins, and you face southerly, the species' richness drops, and hickories, oaks, and pines predominate. But even here, you're liable to name more kinds of trees than you could find in all of Europe.

So what besides God's plenty indicates you're standing in an Appalachian cove forest? One source suggests silverbell and basswood. Popular in garden circles, silverbell (*Halesia* ssp.) is more easily spotted in someone's yard than in the woods except when it offers its white, bell-shaped flowers to the observant. But the basswood, linden, or lime tree is noteworthy by any name—one hundred feet or taller and three to four thick. Bees adore *Tilia americana;* we humans adore their basswood honey, which, if you cannot find, you may approximate by licking the aphid honeydew that often coats basswood's lower branches. Goethe buried young Werther under a basswood, Proust remembered his childhood upon dipping a petite madeleine in a cup of *Tilia* blossom tisane, and Coleridge sulking in his lime-tree bower describes how "Pale beneath the blaze / Hung the transparent foliage; and I watch'd / Some broad and sunny leaf, and lov'd to see / The shadow of the leaf and stem above / Dappling its sunshine!" The Greeks believed that the dying Baucis and Philemon were turned into a basswood and an oak, and these two still provide companionate shade to one walking an Appalachian cove.

Striped maples, basswood, Fraser fir, and many a plant will grow elsewhere beside their native ranges. How else would we ever have a garden? And many an indicator plant indicates not its native habitat but the hand of man. You'll find large sweetgum (*Liquidambar styraciflua*) in many a Lexington yard, but nary one grown from a seed. A glance at the USDA's Plants Database page will show you that the Blue Ridge and Alleghenies form potent barriers to the tree's invasion of the Shenandoah, even though it's native everywhere in eastern and spottily in southwest Virginia. Such transplants, often doomed to die without descendants, betray our disturbing presence. Those that live on long after whoever planted them has gone become a kind of indicator species, three of which—periwinkle, daffodil, and yucca—reveal where cabins stood in what is now wood again. Women who sought a bit of beauty traded one with another plants as tough as the lives they lived. No doubt they planted others, but this tough trinity is all that has lasted the three to four generations since those who planted them last closed the doors to home and vanished. Although House Mountain's people left sometime around World War II and their cabins long burned or rotted into loam, the daffodils—invariably double blooms locals call "scrambled eggs"—still crowd their chimneys' bases, yucca's creamy candles light the gloom, and purple periwinkle shrouds what once was yard. Should you come spelunking for the past, build a fire for old times' sake on hearths that once upon a long time gone gave warmth and cheer through the winter. And if no spring or creek runs nearby, clamber over wind throw and tree fall to find the hand-dug well sunk through twenty feet of rocky till to where water flows, and tie a cup to string and sit and fish in summer for a gift colder and sweeter than any bought in town.

# A MOVING GROVE

Having read too many poems and watched too many movies, I worry when the forest's trees begin to move. Lacie pooh-poohs such fear, pretending to hear in the creak and groan of trunk and tree the stress of wind upon insentient vegetable and not the slow reaching of the thing within that watches all that pass and, tempted—by what? odor, smell, tread, view—, would us embrace, enclose, and trap within its woody arms. Aeneas remembered when a tree whose roots and branches he broke to burn a sacrifice oozed black blood and told the horrified Trojan how it had once been Polydorus, murdered and turned to wood, whom Aeneas placated with fresh funeral rites and earth heaped high to bury the unquiet spirit. Dante encountered in a hellish wood of knots and gnarls and poison thorns a tree that, plucked, "As out of a green brand, that is on fire / At one of the ends, and from the other drips / And hisses with the wind that is escaping" both words and blood— the damned soul of one who killed himself, the aptly named Pier delle Vigne, chancellor to Holy Roman Emperor Frederick II.

No lesser light than Charles Darwin lamented that, in researching *The Power of Movement in Plants*, "we have killed or badly injured a multitude of plants." They died to demonstrate what many long suspected, that plants move. Darwin put the study on scientific footing by affixing slivers of glass to leaves and stems and letting them trace on paper as they moved circles, ellipses, and zigzags. Among those he studied are the *Camellia japonica*, whose leaves revolve in circles in the morning, fall in the afternoon, and rise and fall again at night; dahlia leaves that "undergo a daily periodic movement, sinking during the day and rising at night"; carnations (*Dianthus caryophyllus*) that endure a "highly tortuous and zigzag course, together with some loops"; a passion flower (*Passiflora gracilis*) whose "young leaves sleep by their blades hanging vertically downwards," although a solicitous Darwin notes, "From

82

some unknown cause the leaves do not always sleep properly"; and young oaks (*Quercus* sp.) that move in "irregular ovals or ellipses." Both Venus fly traps (*Dionaea muscipula*) and sundews (*Drosera* sp.) fascinated Darwin, who dreamed all manner of ways to test these plants' remarkable power to move quickly.

Venus fly traps are rare in the wild but *Drosera* surprisingly common if you know where to look. I remember lying on the ground as a child watching them, happily sacrificing ant after fly to their glistening tentacles, and have since come upon them in as an adult in bog after bog. Even more common is *Oxalis*. Though Darwin studied a different species, you perhaps have seen for yourself in the purple-leaved *Oxalis triangularis* your mother gave you when she divided hers its nightly closing of its leaves. Though folded, they do not sleep quietly, says Darwin, who observed that "the leaves after they had gone to sleep, were found to be in constant movement," dreaming of what Demeter only knows. In plants this is nyctitropism, or night-turning, in humans a sign of old age. Native to Brazil, *O. triangularis* is too tender to survive Virginia's winters. But a tougher cousin, yellow wood sorrel (*O. stricta*), grows happily throughout much of North America, its tart flavor a welcome treat on hikes.

Of course, we all know plants move. The sunflower famously tracks the sun, although the heavy-headed monsters Lacie grows soon give up twisting and hang their heads in yellow shame. Some plants, though, will move to the touch of mere mortals like ourselves. Thigmotropism—from the Greek for touch, *thigma*—the educated call it. The South American *Mimosa pudica* (from Latin: *pudica,* shy or bashful) lives up to its common name, sensitive plant, by closing rapidly. Too sensitive to survive the rigors of our winters, its place is taken by our native sensitive brier, which comes in several species. The Eastern sensitive brier (*Mimosa microphylla*) looks like a miniaturized mimosa tree turned into a vine. Spottily distributed in Virginia, it's more common elsewhere, where I've played God with its coward leaves. Partridge pea (*Chamaecrista* ssp.) are easier to find, resembling miniaturized mimosas with pea-like yellow flowers. Although these too fold when nudged or picked, I've found that some are braver than others and have, like Darwin, killed or badly injured a multitude looking for the bashful.

Equally sensitive are some plants' seeds. Orange (*Impatiens capensis*) and yellow jewel weeds (*I. pallida*) take both their scientific—*Impatiens*—and common names—touch-me-not—from their ripe seeds' habit of exploding outward when touched, an apparent adaptation that encourages the spread of these cousins to the less leggy African and New Guinea Impatiens you buy at the store. While you can test jewel weed's projectile capabilities in summer, you have to wait till later to witness witch hazel's even more impressive fireworks. *Hammemelis virginiana's* wispy yellow appears late in the year when few flowers remain, and it's worth bringing a few indoors if the previous year's seed capsules haven't yet opened. When mature and dry enough, they explode with a pop loud enough to hear, throwing their seed across room or mountain edge, and so going walkabout.

Okay, okay. Plants don't really walk, you say. And yet, Mark's gospel records that, upon being cured by Jesus's spit, a blind man said, "I see men as trees, walking." Oh, tree, take up your roots and follow me. And some have. Floridians term mangroves "walking trees" because they seem, on their stilt-like roots, to be about to take a hike. Central and South Americans call the seventy-five-foot tall *Socratea exorrhiza* the "walking palm" because its stilt-like roots—taller than a man—seem to be ready to carry it off. And, according to the botanist John Bodley, when knocked over, the palm grows new roots, righting itself and so moving, albeit slowly, through the forest. We too have slow-moving plants that layer their branches in leaf mulch and propagate new selves and march across our flower beds and forests. Toppled trees whose roots still work will sometimes reroot themselves and so move a trunk's worth through the woods. Tropical walking irises of the genus *Neomarica* take their common name from their habit of forming new plantlets when flowering, bending over and letting them root as they "walk" the forest. Here in Virginia the long-leafed walking fern (*Asplenium rhizophyllum*) forms new plants when its leaf tip touches ground, tiptoeing down the limestone ledges flanking the creeks it favors.

Such feeble vegetables bore Edward, who has long wished, as have I, to see an Ent, like those in *Lord of the Rings*. The tree people's name came from the Old English Tolkein studied, *ent* meaning giant, as in Beowulf's "eald enta geweorc," the old work of giants. Although Edward and I stood awed beneath the redwoods of California, they are too gigantic to be Ents, Burke's distinction between the sublime and the beautiful pertinent when gazing on such a volume of living vegetable. What we desire is something human scaled, like the eald enta *Quercus* that watch us as we walk the ridges of their world. Surprise a thick-barked veteran of centuries of war with man and sky gods, you can half-hear in its leaves' muttered whisperings the something Tolkein wrote of, "that grew in the ground—asleep, you might say, or just feeling itself as something between root-tip and leaf-tip, between deep earth and sky had suddenly waked up, and was considering you with the same slow care that it had given to its own inside affairs for endless years." Should the evening catch us miles from the car, we walk a little faster, suppressing thoughts of timber and paper products, rendering respect to the ancient ones, pressing hand and lip to mossed and lichened trunks wiser far and longer lived than we. Then I remember the one who told Macbeth, "As I did stand my watch upon the hill, / I look'd toward Birnam, and anon, methought, / The wood began to move." The creak and groan of trunk and limb accompany our ever-faster walk as darkness gathers. We regret the trees we've cut and bled throughout the years and, breaking through the forest's edge, behold a grove of new-sprouted oaks that have escaped the teeth of cow and deer encroaching on the field, the young of eald enta *Quercus* watching from the woods' edge. Below them, along the creek, sycamores as eald and enta as an oak eye us, one of their number years gone having, I once thought, fallen in a storm. But now I perceive, as through a glass darkly, this was no fall; so set in ways so long it half-forgot the art of walking, this *Plantanus occidentalis* ripped its roots from smothering earth and crawls now uphill toward the descending *Quercus*. The sassafras whose fencerow roots I cut in spring for tea I now know all one tree grown horizontal, the black locust groves whose flowers intoxicate me in the spring the heaven-reaching fingers of a circling eald enta. Believe me, the woods, they move.

# A TANGLED BANK

Legend has it that Charles Darwin politely declined Karl Marx's offer to dedicate *Das Kapital* to him. Marx never actually offered, but he did send Darwin a copy of his book (whose uncut pages reveal Darwin never read it all) and wrote a friend two years after *On the Origin of Species* was published that "Darwin's work is most important and suits my purpose in that it provides a basis in natural science for the historical class struggle." The union of evolution and revolution, economics and biology that Marx celebrates has not been altogether happy. And economics is but one in a long line of husbands biology has resorted to to support sometimes insupportable assumptions.

Take the bee, a metaphor for whatever theory you wish to advance. The Roman poet Virgil describes in his *Georgics* their chastity: "they don't indulge in sexual union, or lazily relax / their bodies in love, or produce young in labor, / but collect their children in their mouths themselves from leaves, / and sweet herbs, provide a new leader and tiny citizens themselves, / and remake their palaces and waxen kingdoms." Which perpetual virginity permitted medieval nuns to keep bees without fear of threatening their own chaste state. And note, bees dwell in "palaces and waxen KINGdoms." Not until the microscope did we learn that the king bee was, ahem, a queen. Shakespeare's Ulysses says in *Troilus and Cressida* that "For so work honey-bees, / Creatures that by a rule in nature teach / The act of order to a peopled kingdom. / They have a king and officers of sorts"—magistrates, merchants, soldiers, masons, "civil citizens kneading up the honey," mechanics, and justices "Delivering o'er to executors pale / The lazy yawning drone." All in all, the perfect parallel to Renaissance notions of polity.

By Milton's day, the king had lost to his wife his crown. But rather than a gynocracy, in *Paradise Lost*, God creates "the Female Bee that feeds her Husband Drone / Deliciously, and builds her waxen Cells / With Honey stor'd," the perfect model of the busy bee. And an object lesson for wives. So popular did busy bees become that the Mormon Church adopted their hive as symbol, the *Deseret News* (Oct. 11, 1881) explaining, "The hive and honey bees form our communal coat of arms . . . . It is a significant representation of the industry, harmony, order and frugality of the people, and of the sweet results of their toil, union and intelligent cooperation." Seven years later, Anna Quillin drew another lesson: "I presume the bees are believers in 'woman's suffrage' and 'woman's rights' for they are always governed by a queen, and it used to be asserted that the females did all the work." Sylvia Plath found yet another message in the queen's unceasing toil in *The Bee-Meeting:* "The villagers open the chambers, they are hunting the queen. / Is she hiding, is she eating honey? She is very clever. / She is old, old, old, she must live another year, and she knows it." From king to drudge is quite a fall.

Not everyone was as enamored of the bees' "industry, harmony, order and frugality" as were the Mormons. Bernard Mandeville attacked frugality in his 1714 *Fable of the Bees,* arguing that economic prosperity comes not from frugality but from excess. "Bare Virtue can't make Nations live / In Splendor; they, that would revive / A Golden Age, must be as free, / For Acorns, as for Honesty." Mandeville's observation that, with selfish bees, "Thus every Part was full of Vice, / Yet the whole Mass a Paradise" anticipates Adam Smith's "invisible hand" in which personal greed improves society as a whole. Asserting in *Bumblebee Economics* thirty years ago that "'economy' is almost synonymous with 'frugality,'" Bernard Heindrich analyzes bumblebees in terms of "revenues," "direct costs," "gross profit," "overhead costs," "net profit available to reinvest." Stung by Michel Rothschild's use of *Bumblebee Economics* as "a scientific justification for capitalism as espoused by political free-market advocates," Heindrich—whose book is endlessly fascinating—observes that "what applies to bees does not necessarily apply to us."

Economists are not alone in appropriating other disciplines' work; like Heindrich, ecologists use economics to explain nature. Patrick C. Kangas and Paul G. Risser's 1979 "Species Packing in the Fast-Food Restaurant Guild" answers the simple question "Why are there so many kinds of FFRs [fast-food restaurants] along the strip and how do they coexist as possible competitors?" by applying the ecological concept of species packing—the maximum number of competing species that can coexist in a single community. I had never before thought of a fast-food strip "as being composed of sessile consumers which filter out resources flowing past their structure." Never supposed it like "an oyster reef or a floodplain forest" or "web spinning spiders, net-spinning riffle insect larvae in streams and inflorescences which compete for pollinators." To help "particulate organic matter" such as myself visualize the speciation of FFRs, Kangas and Risser include a chart displaying "a hypothetical phylogeny" that supposes an "ancestral drive-in generalist" whose descendants diversified into today's large and small hamburger specialists, chicken specialists, hot dog specialists, roast beef specialists, and Mexican and seafood specialists.

A University of Vermont webpage converts the notion that Walmart kills off competitors into a mix of economics, ecology, and politics, observing that "[t]he 'Walmart' community has a much simpler structure than the one with a more even distribution of species (this characteristic is often called 'eveness' and is an important aspect of biodiversity). There has been much debate over the relative stability attributed to these different community structures, with many suggesting that the more even or equitable community structure is more diverse and more stable.... What would Adam Smith say?"

The notion that complex and redundant ecosystems are more stable than simpler ones is taken from, of all things, information theory. The mathematician and cryptographer Claude Shannon basically created information theory in a 1948 paper that tried to figure out a way to measure the uncertainty in a message, its entropy, as it were. The Shannon diversity index will give you a number representing just how garbled what you want to say is:

$$H^1 = - \sum_{i=1}^{R} p_i \ln p_i$$

A formula that, to me, has an astronomically high level of entropic incommunicability. Ecologists took this measure of textual uncertainty and used it to determine the diversity index of a community—how many different kinds of species there are on a mud flat, say, or in a strip mall. Faced with economics, physics, ecology, mathematics, information theory, politics, and who knows what all busily interbreeding in the basement, someone as mathematically illiterate as myself must to a trigonometry class to learn how Shannon's index might measure the entropy in Darwin's conclusion to *On the Origin of Species* that "It is interesting to contemplate a tangled bank, clothed with many plants of many kinds, with birds singing on the bushes, with various insects flitting about, and with worms crawling through the damp earth, and to re-flect that these elaborately constructed forms, so different from each other, and dependent upon each other in so complex a manner, have all been pro-duced by laws acting around us. These laws, taken in the largest sense, being Growth with Reproduction; Inheritance which is almost implied by repro-duction; Variability from the indirect and direct action of the external condi-tions of life, and from use and disuse; a Ratio of Increase so high as to lead to a Struggle for Life, and as a consequence to Natural Selection, entailing Divergence of Character and the Extinction of less-improved forms. Thus, from the war of nature, from famine and death, the most exalted object which we are capable of conceiving, namely the production of the higher animals, directly follows. There is grandeur in this view of life, with its several powers, having been originally breathed by the Creator into a few forms or into one; and that, whilst this planet has gone circling on according to the fixed law of gravity, from so simple a beginning endless forms most beautiful and most wonderful have been, and are being evolved."

*On the Origin of Species'* last word is "evolved," the first edition's only use of the word we've come to associate with things Darwinian. That the same sentence also contains "the Creator"—sometimes cited as proof of a religios-ity few but the born-again believe Darwin possessed—only complicates the work's diversity index. As do the "laws" Darwin refers to—a metaphor taken literally by Social Darwinists desiring to find in nature justification for hu-man laws regarding who can marry whom. The poet William Blake warned of the triumph of such abstraction in his prophetic *Song of Los:* "Thus the terrible race of Los & Enitharmon gave /Laws & Religions to the sons of Har binding them more / And more to Earth: closing and restraining: / Till a Phi-losophy of Five senses was complete / Urizen wept & gave it into the hands of Newton & Locke." Los and loss, Urizen and your reason—Blake has little use for so-called laws that bind us. The *Oxford English Dictionary* explains that "The 'laws of nature', by those who first used the term in this sense, were viewed as commands imposed by the Deity upon matter, and even writers who do not accept this view often speak of them as 'obeyed' by the phenom-ena, or as agents by which the phenomena are produced." Locke and Newton, Blake thought, sought to imprison our free nature in imperfect "laws" that prescribed what might and proscribed what might not be done—Locke's so-cial contract a secularization of the Christian notion of Adam's original sin binding all his descendants forever. Worse was to follow—naïve capitalists to this day pretend that an economic system but three hundred years old that breeds money by interest—once thought blasphemy by Christians—is "natural," justifying what they practice in bank vaults by appeals to Darwin's "tangled bank." But Newton's "fixed law of gravity" is just that—fixed, not some catch-as-catch-can capitalist or communist conundrum cobbled together to facilitate Paul's robbing Peter of his pennies or rubles. Behind Darwin's "war of nature, . . . famine and death" lies Thomas Malthus's dismal vision of famine, disease, and war as checks on human population growth and Thomas Hobbes's natural state of man, a *"bellum omnium contra omnes,"* war of all against all—"continual fear, and danger of violent death; and the life of man solitary, poor, nasty, brutish, and short." Hobbes's solution? "A common power to keep them all in awe,"—Leviathan, the state.

And it was the state in Darwin's day that enforced the caste laws of India, which, to echo *On the Origin*'s words, oversaw both the "less-improved forms" and "the higher" Brahmin. Darwin's insights would be twisted into much of the worst bigotries of the modern world, and it is a delicious irony that the *OED* credits him with being the first to label the worker and the soldier ant as member of *castes*. His observations on bees, like Heindrich's, are open to misconstruction by anthropomorphizing improvers of the human lot. When Darwin asks, "If we admire the truly wonderful power of scent by which the males of many insects find their females, can we admire the production for this single purpose of thousands of drones, which are utterly useless to the community for any other end, and which are ultimately slaughtered by their industrious and sterile sisters?," both my inner bigot and my slug stir, the one to impose his hates on others, the other in fear of imposition.

# SWEET AND SOUR

Limestones laid down in seas gone long before the dinosaur make the Valley of Virginia the marvel that it is, breadbasket to the Confederacy and so rich a pasture that Rockbridge County has more cows than people. Geologists are priests of the past, apprentices in the mystic art of reading in the convoluted entrails of the earth what has been and may yet be again. It's worth the promise of a six-pack to convince one to read a cliff face as you might a fairy tale and stand entranced by once upon-a-times and once-there-weres until you fly Virginia for Laurentia and continents unheard of, the Atlantic shrinks and disappears, appears and disappears again and yet again, mountains undreampt of rise and fall like morning fog, and back, back, back you're taken through time's tunnel to the shores of shallow seas that stretch westward to where, in a future not yet guessed, an Iowa may one day stand. With but a ten-power magnifying glass, the past's pope pronounces this stack of grayish stripes a dessicated mudflat whose scales of cracked mud you can still trace in stone. I lick the grainy stuff and imagine me a worm feeding on algaed stromatolites that grow like cabbage heads along this cliff that once was seashore. And think perhaps I taste the salt of vanished seas.

Anyone can read a poem. But to read mountains—now there's a skill worth taking classes in. The hammers of countless rockhounds scenting tracks of stone have built a library's worth of wisdom the government has mapped in quadrants so colorful they're worth the framing so, house bound, you can look on the land you stand upon and understand, if but dimly, what has made it. A mere two hundred years ago you had to take Moses at his word and suppose the flood of floods carved what you saw. No more, and now worlds more marvelous than even New Jerusalem rise before our eyes. You can drive U.S. 60 east to west across Rockbridge County in less than an hour, dropping from the Blue Ridge's granite rampart down through the resistant quartzite sugar loafs to the Valley's yielding limestones and dolomites and up again to North Mountain's harder sandstones and know nothing but that the land rises, falls, and rises. Or disciple yourself to the apostles of the past and learn that where we live today was predestinated by a gospel scratched in stone.

Lime so enriches soils that you can buy it for your backyard garden. Imagine, then, the settlers' joy on discovering that Shenandoah's grasses grew on limestone. So enamored were they of the stone that a happy someone, surprised to find it where all else was sandstone, named his piece of paradise Rich Patch, a name that stands to this day for a narrow grassy pastureland enclosed by piney ridges. The best land soon sewn up in deeds of ownership, latecomers settled higher and higher up the hollows, hoping against hope the sandy till they found but a summer's blanket worth that, thrown off, would reveal gray sheets of lime. Philip Clayton's perch aside House Mountain is one such hope, its cabins built where long-gone dreamers dreamed the stony rubble that would break a plow might hide a sweeter secret. But that till's twenty acidic feet in places, and Philip's is the last cabin still inhabited that high up the mountain side. Sugar Creek's half-hearted effort to wear the mountain down has dug here and there to limestone bedrock. But higher up, the creek ends, and sandy, acidic soil cast down by ice age upon ice age broke even the backs of settlers determined enough to build where both lime and horizontal lines were fantasies.

Abandoned, logged over, sold out to the government, the sandstone ridges flanking the Shenandoah are where our forests are because farmers could not make a go of it upon their flinty slopes. Others tried, when dreams of iron foundries filled promoters' ledger books with dollar signs. Geologists have puzzled out the complicated chemistry that led iron-rich waters percolating through the tilted sandstones to drop their burdens at the mountains' feet for miners to dig up and smelt in furnaces whose chimneys stand today in forests grown back from when charcoalers felled the trees for miles around to feed the furnaces' voracious appetites for wood.

The Civil War and cheaper, more abundant Great Lakes iron ore ordained that the Valley would remain a rural paradise. Should you hike St. Mary's Wilderness, you'll walk where trains once steamed and miners delved and camp among their ruin. St. Mary's stream is cold enough to harbor native trout, remnants of the colder Pleistocene. Just as the hollows acid soils ran out would-be farmers, now St. Mary's waters doom her trout. Sprung from bedrock, the water, hovering on the edge of being too acidic, has become poison to its fish from the incessant fall of rain made acid by midwestern power plants. Desperate Forest Service scientists once dropped helicopter-loads of crushed Valley limestone in a vain attempt to stave off disaster. Fish while you can; the trout, like the farmers and miners, are doomed.

High up these hollows, where the soil thins to an intermittent skin stretched patchily over bare bedrock, live acid-loving communities of plants. Heath, botanists call this carpet of blueberry, huckleberry, and winterberry begging for foragers willing to tackle the higher-reaching laurel and rhododendron through which rises a sparse forest of Virginia pine, maple, and birch. Hiking up St. Mary's trails in August in search of berries, I've come across rivals armed with revolvers against, they claim, snake and bear. Rattlers are rumored to love these rocky wastes, but I've never scared up one. And the bear I met atop North Mountain didn't look as if a pistol would deter it. Carrying down a backpack full of blueberries for cobbler and pie, I've worried more that I might be robbed by two-legged than by four-legged or legless varmits.

Many a time I've dreamt the reverse dream of Valley settlers, of how I might bring down the mountains' acid soils and grow an easy harvest in my own backyard, in whose sweet soil blueberry and wintergreen, azalea and rhododendron turn up their noses, wilt, and die. Vinegar, Lacie once advised, and, can I but remember, I douse the soil with bottled vinegar the rain soon dilutes to useless quantities. Sometimes I think I should sit beside my plants and let the acid of gossip drip down on them. But don't, for fear I might turn this paradise into a desert.

Desperate for a taste of winterberry, I take me to the mountain tops, where 200 million years of weather have carved sandstone castles atop whose turrets I can sit and look down on the workaday world of the busy bodied and eat, one by delicious one, red berried treats I steal from none but chipmunks.

# ROUND AND ROUND
# THE MULBERRY BUSH

Remember marveling as a child at the sassafras's many forms of leaf? The normal, grade-school-leaf-looking variation, the right- and left-handed mittens, the double-thumbed extravagance? Collecting and pressing these mysteries between a book's leaves and coming on them decades later, brown and brittle, though their memory be as green as sassafras in spring? And how no adult knew the why or wherefore of this jeu d'esprit of God's exuberance?

Still no adult knows why *Sassafras albidum* was picked for peculiarity. Or why the paper mulberry (*Broussonetia papyrifera*) shares an even greater propensity for polydactility in summertime—knitting mittens with none to five fingers. Or why both our native red mulberry (*Morus rubra*) and the imported white mulberry (*Morus alba*) show a similar plasticity of form that the botanical call heterophylly. Propensity for shapeshifting is thought to be genetically hardwired—something even inexperts such as myself can see in sassafras's and mulberry's continued insistence on the trait. But more observant eyes have sought to find as well in heterophylly a response to the environment, some saying sassafras leaves lobe more if in the sun or at the ends of branches, while shaded sassafras tend to be entire. Others have observed that trees whose branches are multilayered permit light to filter branch to branch by staggering or lobing leaves to let light through. Botanists with more tools than you and I have discovered a subtler heterophylly than eyes can see—sun-blessed leaves grow thicker and more massive and possess more stomata than do shade leaves.

Mature leaves, like mature humans, tend to more rotund shapes than their juvenile counterparts. Both the birch (*Betula* ssp) and the trembling aspen (*Populus tremuloides*) replace lengthier youthful leaves with maturity's rotundity. Your backyard ivy's (*Hedera helix*) mature leaves are larger and less lobed than earlier ones. The older holly (*Ilex* ssp) exchanges its youthful leaves, as spiked as any Goth's haircut, for a less prickly state. Red cedar (*Juniperis virginiana*) too exchanges youthful spikeness for an adult flat spray just made for hanging Christmas ornaments. Striped (*Acer pensylvanicum*) and red maple (*A. rubrum*) both exchange the pronounced lobes of youth for blobbier, less-lobed later leaves.

Aquatic plants are even more famous in botany circles for their heterophylly. The northern-growing common mare's tail (*Hippuris vulgaris*) grows longer leaves submerged. The wider distributed water crowfoot's—aquatic buttercup (*Ranunculus* ssp)—submerged leaves are threadlike, its floating ones rounder, as I can verify by wandering the banks of many a Valley stream and playing with Edward butter-on-your-chin with its shining flowers. The yellow pond lilies—spatterdock (*Nuphar advena*)—that grow just outside Lexington, despite their owner's various attempts to poison them, continue their aquatically inspired plasticity, growing heart-shaped to round leaves that may be under, on, or out of water. Their yellow-as-the-sun flowers form woody seeds pods well worth a surreptitious trespassing to steal, while their algae-covered submerged stems are an aquarium delight of microorganisms. As variable as arrowheads we find in plowed fields are the arrowheads (*Sagittaria*) of Valley ponds, whose leaves shape change from air to water. As do those of the bane of ponds, the pondweeds, whose strap-like floating leaves far outdo their aerial siblings and, brought in upon the feet of goose and duck, choke many a Valley farm pond.

Heterophylly provided Victorian botanists a tool for explaining how plants came to possess such a marvelous variety of differingly shaped parts. Such obvious differences between aerial and submerged leaves seemed perfect proof of Lamarck's theory of inherited characteristics. Yet even after Darwin's competing theory had beaten out Lamarck and married Mendel's genetic inheritance model, botanists had to wait until the notion of phenotype arose—the idea that the environment can influence the way a gene expresses itself—to satisfactorily explain the watery buttercup's two types of leaves as the temporary influence of environment, not genetic mutation. Others had found in heterophylly confirmation of the German biologist and visionary Ernst Haeckel's now discredited biogenetic law that ontogeny recapitulates phylogeny. Which is that in the development of the individual—especially in embryo—we see repeated the evolution of the species. That tree leaves seem to change shape as they mature and then revert in senescence to infancy seemed visual proof of what otherwise you had to seek in fetuses preserved in formaldehyde.

The youthful Goethe in search of a mystic vegetable *Bauplan* exclaimed, "All is leaf." And he was right—the most complicated flower there is is, finally, but a bunch of leaves arranged in whorls. Even I can see that green sepals and colored petals are modified leaves. And, once told, I can see that a violet's various petals are a sophistication of the magnolia's plainer plan, the pitcher plant's pitcher and Jack-in-the-pulpit's pulpit both but leaves wrapped round themselves. The acorn lifts its cotyledeons heavenward in open-leaved appreciation of the sun, and an entire tree springs from what was leaf. Goethe was right—all, all is leaf, and every plant a book writ in mystic hieroglyphs awaiting its Champollion.

Read in the daisy's and sunflower's packed flowerheads the Chaldean cuneiform of Fibonacci numerology, in the trillium's adoration of the Trinity a Christian calligraphy wound round a plant stem, in the rose's perfumed petals the same whorl of leaves my son, back-stripping a stalk of privet, bouqueted in his fingers and threw me, asking, Want a flower, Daddy—all this world's angiosperms whirling round and round in held-to-heaven floral whorls in which botanizing dance instructors have divined a fourfold pattern of perfection: the whorl of sepals a calyx on which dance the corolla, the petals' whorl, from the simplicity of splayed magnolias to the rose's petalled profusion, the male androecium's measured whorl of stamina whose anthers dance, dance, dance around the female gynocium's whirling stigma, style, and ovary.

The mystic music only plants can hear drives the Indian cucumber, Joe Pye, lily, and mare's tail to dosey-doe even their leaves in whorls that climb stems dead by autumn's end. Pines hear a more enduring measure mere mammals are too deaf to sense, and year by year they thrust their branches skyward in whorls that are the turning record of their lives. Hold a spruce tip in your hand and see the whorl that built the tree you stand beside. Or stand beneath a sixty-foot-tall conifer and see rise above you like a staircase to the sun whorl after whorl of branch, and count their measured growth through time. The same turn, turn, turn the trees intoned in your childhood when you climbed round and round their trunks from whorl to whorl until you rose above the world of everyday and saw beneath you the globe itself turning round and round the sun. Though you and I, old now, are deaf, the trees still sing their turns, and some, so taken with desire to whorl, dervish on themselves their grain as if they would escape the hold of earth and dance the rites of spring. Stunned by a great rooted chestnut blossoming in tasseled whorls, neither we nor Yeats can tell the dancer from the dance.

# A FOREST FIELD

🌿 Wander the woods hereabouts, and you soon learn that what now is forest once was field. And that fields and their maintenance have permanently marked the land. Indians, who had all but vanished from the Valley by the time the white man made it over the Blue Ridge, created with fire grasslands that prophesied what Europeans would find west of the Mississippi. Not that everyone appreciated their desire to turn the Valley into a giant pasture; the West Virginian Hu Maxwell complained that "The Indian is by nature an incendiary, and forest burning was the Virginia Indian's besetting sin.... [I]f the discovery of America had been postponed five hundred years, Virginia would have been pasture land or desert." No doubt Maxwell's job as a timberman colored his appreciation of treeless expanses out of which arose unburned islands of trees remembered in today's place names: Brushy Hill, Timber Ridge, Walnut Grove. The white man continued burning what he could and harvesting the rest.

Driving the Shenandoah's back roads, we forget that the cattle-filled fields we love are late additions, that much of what is pasture now was wood, then fields of wheat or corn. It was in Rockbridge County, after all, that Cyrus McCormick invented his revolutionary reaper—designed to harvest the Valley wheat fields. The Midwest's larger fields made him a wealthy man and broke the breadbasket that the Valley was until after the Civil War. And Prohibition made superfluous the corn once used to manufacture whiskey, sealing the fields' fate. Dairies still grow cow corn, but most Valley fields are now devoted to hay or pasturage.

It's instructive driving by fields along the interstate to speculate on which is what. Even in winter, last year's stubble of corn stalks betrays a field's purpose, especially now that genetically modified seed has made no till preferred agricultural practice. Bedrock so chokes some fields that neither plow nor tractor can enter them, and these you know are pasture. Cattle won't eat everything that grows in a field, and you can guess by one's ragged looks it's pasture. A hay field, on the other hand, will be a sea of waving grass, spring green or amber waved. If mown, it will rival your lawn in curb appeal. And, of course, if dotted with the round bales that have replaced the square bales of my youth which replaced the haystacks of earlier generations, it's not for grazing. Many a Valley boy once learned by heaving square bales onto a flat-bed just how heavy grass can be. Where once a farmer paid good money for boys to get an upper-body workout, today they must to a gym and themselves pay others for the privilege of building muscle.

If you don't want to lose your cattle, you must fence them. Not that the law requires you too. In Rockbridge, the law says you're responsible for fencing critters out, not in, and many a fence is so feeble it falls when a mosquito lands upon a strand of wire. Iowa State extension agents figured the costs of various fences in 2005: a quarter mile of woven wire—1,320 feet, about enough to fence a football field—would set you back nearly $2,000, five strands of barbed wire $1,614, eight strands of high tensile $1,483, five of the same electrified $927, and a single strand of electrified polywire $240. Which explains why you now see so many dinky electric fences in the county. What, though, to do with the old fence? Some just leave it to become a prop for cedar, cherry, honeysuckle, and multiflora rose. Root it out, and you've got to dispose of it, which is why God made sinkholes. The same Iowa agents guessed that a fence is good for only twenty years, though there's plenty a Rockbridge fence that looks like it's lasted through more presidents than that. There are folk that can date a fence by the kind of wire it uses, and you can visit barbed-wire museums, if you've a mind to. It's just as easy, though, to cut a cedar volunteering through a fence and count the rings to get a rough idea of how long ago a farmer ran the wire. Or guess by how deeply buried in the trunk it is when he round wound the wire loop now killing a tree.

Fencing has had its share of fads. Chain link was being manufactured by the 1840s. Woven wire, which is a kind of knotted chain link, followed. As did barbed wire, two strands of wire twisted round each other with barbs inserted every couple of inches—so inexpensive it fenced the West. Electric fencing became feasible in the 1950s. And high tensile came in in the 1970s. Before these various metal fences beat them out, wooden fences—too expensive to build and maintain over long distances—rung our country's farmyards. Valley farms still use four-boarders for their feed lots, and fancy horse farms fancy wood over synthetic or wire. But before wire, enterprising farmers turned to nature, dragooning thorn upon thorn bush in their quest to build a fence horse high, hog tight, bull strong. Which is how the Osage orange (*Maclura pomifera*) came from Oklahoma on the range to the East Coast, visionaries' eyes sparkling at the prospect of thorny barriers to wandering animals. Named *bois d'arc* by French explorers impressed by its use in Indian bows, millions of bodark, as the Ozarks knew them, sold until upstaged by barbed wire in the 1870s. Though abandoned, bodark refused to die and now grows, forgot by all but children who call it the monkey brain tree in honor of its cerebellum-like fruit. An even shorter-lived craze was that for the multiflora rose, extolled in the middle of the twentieth century by government agents as the perfect natural fence, who forgot that, spread by birds, *Rosa multiflora* would turn fields, not fencerows, into Sleeping Beauty's impenetrable forest.

Before wire and thorns and boards, there were sticks and stones. Crude fences made of stumps and branches pulled from newly cleared fields came first. A more sophisticated version, the Virginia worm or snake fence, versions of which the Park Service erects on famous farms and battlefields, was a zigzag of interlocking split rails. And rock walls rose where the land was filled with portable stone. Iconic New England thanks to the glaciers, they made such good neighbors despite Robert Frost's low opinion of them that hundreds of thousands still stand where settlers built them to clear and then delineate their fields. The Valley, being limestone, had only rocks forgotten by the river, and these, too few for fences, the settlers dumped in sinkholes or piled in cairns. But Piedmont Virginia's granite bedrock and the mountain's sandstone ridges provide rock aplenty for walls that now smell of money in Loudoun and Albemarle counties and sink forgotten by all but chipmunks and wood wanderers in less populated mountain coves.

Sit on one of these for a decade or so and learn how little Frost knew of arborculture. "He is all pine and I am apple orchard. / My apple trees will never get across / And eat the cones under his pines, I tell him. / He only says, 'Good fences make good neighbors'." Slower than cattle though trees are, they move, and pine and maple press in upon the apple orchard high up on House Mountain's saddle, and apple trees gone rogue raise their branches on Cole Mountain. Frost blames frost-heave and rabbit-hunters for his walls fall, and he might have added the rooting out of rock by root and trunk, which can heave a wall to ruin as slowly as a tree grows, sending its rounded cobbles dispersing through the litter like high domed tortoises.

Before you leave, take what you know of Valley fields and look about. Was what is forest once pasture? Hay? Or plowed? Archeologists can follow the track of a plow through the soil it cut, and medieval farmers plowed so deep and long their ridges and furrows still stand in English meadows. Virginia's farmers tilled less deeply, it would seem, and I know but two forests in which furrows are still defined enough to see. Cattle hooves cut tracks through hillside fields, and these last longer in the woods than a shallow plow's track. Field size can hint at whether walls or cairns mean tilling or haying: flanking Brown Mountain Creek are stone-walled gardens too small to hay and so chock-a-block with rock that, even after fencing, they had to pile the rest in cairns. On House Mountain's flanks stone walls still stand, and one small walled enclosure with no trace of stone must have been plowed and planted. Uphill, a larger enclosure still studded with occasional stone suggests a pasture lot.

Of course, you think, walls such as these could not have kept much out. Perhaps a less-than-nimble cow or horse uninclined to jump. A wooden fretwork that rose another foot or more above the stone left no more trace three generations gone than did a wooden shed and outhouse. But stone endures. I've hiked past dry stone walls in Cornwall they say stood much the same when the Normans came, when the Romans left, when Phoenicians mining tin first landed. The men and women who heaved up Virginia's testaments to dreams of farms in what was wilderness have vanished. But their handiwork in stone endures. Long after you and I and our children's children's children vanish, when perhaps a people worshipping in strange tongues stranger gods possess this land, these low lines, gray and lichen covered, will speak of dreams once as bright as a cobble fresh dug from dirt and raised thigh-high in pursuit of happiness. And, knowing this, regard here where the wall that taught you this has fallen, and ask which character in Mending Wall you are.

# SNOW FALLING ON WATER

🌿 All summer I walk the woods and learn by leaf and scent and taste the trees' names—the long-leafleted walnut's musky smell when crushed, the pig and bitter and shagbark hickories' subtler aroma, sassafras's many-mittened odor of childhood, black birch's dancing leaves and wintergreeny twigs, spicebush's perfumed handkerchief of leaf, and think I'm knowing something. Come winter, though, my knowledge dies with the leaves, and I'm left as bare and naked as the trees. Who still speak, if only I will listen.

In August's cool week, when the first cold front comes south from Canada, clearing and bluing the sky, and you need a summer sweater to walk outside in the morning, the trees too know fall is in the air, that time we name after their great giving up and throwing down of summer's unheeding happiness in a riot of death so beautiful we advertise for tourists to come a thousand miles and watch it bleed out on the Parkway. Clothed in green, even the most tormented tree casts shade in summer and shields us from the sun's truth. Come winter, bare, gnarled, and broken branches betray the struggle that life is for vegetables, for animals, for us, the stark beauty of winter trees life's fist raised against the indifference of death.

Masked by a chlorphyllic carnival of green, carrot-colored carotenoids spend spring and summer absorbing the light energy used in photosynthesis while protecting cells from sunburn—just as they're said to help protect our retinas from UV damage. Eat, then, more carrot, corn, banana, pawpaw, and persimmon if you would not go blind to the beauties of the fall foliage of the beech's and aspen's yellow groves, the colored bouquets ash, hickory, oak, and tulip throw toward the sun, and the sycamore's great-armed embrace of summer's end. As befits a cellular workhorse such as the carotenoid, it triumphs over chlorophyll's collapse in quiet shades of yellow fading into the brown of death when all that's left in leaves is tannin bitter as unsweetened tea.

Not so its gaudier cellmate, the anthocyanin, author of autumn's reds and purples. Anthocyanin? Down I go to the library, where shelves are lined with books whose black and white leaves shed knowledge—anthocyanin, from the Greek *anthos*, flower, and *kyanos*, blue, a water-soluble pigment whose color varies according to pH. Aha, my old friend, the hydrangea, whose bright blues of my childhood in the coast's acidic sands fade to pale pink in the Valley's sweeter soil. But why, plants, do you bother? Some think the earliest on land came up with pigments as a way to protect themselves from the sun's UV rays before enough ozone accumulated to do the job. If so, anthocyanin's been drafted to serve as well in attracting pollinators, explaining the ubiquity of pale blue flowers and your hummingbird feeder's gaudy red. Those naming things call this a pollinator flag, a sort of come-hither, sailor billboard waving a scented handkerchief in the breeze to one and all. Its fall counterpart they call foliar fruit flagging, an alliterative mouthful for dogwood, tupelo, sumac, Virginia creeper, and, yes, poison ivy's leaves' brightening red, presumably to advertise their equivalent of fast food to birds interstating south to warmer climes. Others think anthocyanin's bright colors advertise a tree's health, guiding would-be pests to weaker targets. Yet others imagine they aid in the reabsorption of chlorophyll. But I, I know the reason why. As summer fades, sensing death, the leaves bequeath their sugared chlorophylls to trunk and root, drain off their green and make anthocyanins in a forest-wide bright blaze of rage against the dying of the light.

And why this green rush to death? Botanists believe the difficulty of obtaining water through a cold winter and the burden of replacing bug-bit and wind-whipped leaves make leaf fall happen. Triggered by the lengthening shadows of the sun's departure, deciduous trees prepare for winter by developing cork-like abscission layers (from Latin *abscindere*, to cut off ), weakening the ties that bind stem to twig, until wind and gravity undo the two, and fall begins. Reluctant to divorce, oak and beech leaves linger long into the winter, curled and complaining in the wind. Come spring, though, new life will have its way, and all fall, making room for others.

That new life, in the form of buds, was set by midsummer, according to the National Arboretum. The rest of summer, they say, the leaves devote to manufacturing carbohydrates for the next year's growth. Always, then, a year ahead of me, trees dream themselves a better future as they face, naked, the indifferent fury of mountain winters. Having bit young the fruit of the tree of knowledge, I would know their names by branch and bark and bud. The tree guides try to guide me. Are the branches alternate or opposite, they ask. If alternate, try maple, ash, dogwood or buckeye. If opposite, they list a forest's worth of choices: oaks, hickories, sycamores, birches, on and on until I scream I'm not any more in school and close my book, take up my stick, and turn for salvation to the woods. Forget our names, they say, our uses in the lesser world below. Look and learn. I run a hand against thick furrowed bark, bend near, sniff moss smell and dirt, leaf loam and death, and underneath, but barely known to nostrils deadened by the city, the hint of life, deep buried, drawn in upon itself, holding its breath for spring.

My father always planted trees in fall, to give them time to become acquainted, he said, to their new surroundings. Here, though, in Virginia, Arbor Day falls on the last Friday in April, perhaps because the world's first Arbor Day was Nebraska's, also held in April, when they say they planted a million trees on the prairie. Since then, countries around the world have designated official Arbor Days, whose dates vary from Australia's July (their dead of winter) to beforested Canada, who devotes an entire September week to trees. China demands everyone plant five trees a year or pay a fine. Israel's national holiday, Tu Bishvat (the fifteenth day of the month Shvat), the new year for trees, usually falls in January or February, commemorating Leviticus's command: "And when ye shall come into the land, and shall have planted all manner of trees for food, then ye shall count the fruit thereof as uncircumcised: three years shall it be as uncircumcised unto you: it shall not be eaten of. But in the fourth year all the fruit thereof shall be holy to praise the Lord withal. And in the fifth year shall ye eat of the fruit thereof, that it may yield unto you the increase thereof: I am the Lord your God." Five years to wait before I can circumcise and eat the apples off the tree I planted in my backyard—in September, because, God or no God, Tu Bishvat in Rockbridge is no time for planting anything.

My daddy knew as much. And I'll take his advice over Yahweh's, whose Mideast mountains I've seen in pictures, bare and treeless. Here in the Valley, come Tu Bishvat, you're likely to have snow. And when it comes, I go to the woods, excited as only one who saw his first snowflake at twelve can be. That boy's still inside me somewhere, sticking his tongue out to catch a flake and trying for all he's worth to see if every one's six sided and different from all others. I would be the first to walk the snow's untrodden trail, thinking that I will there discover something new. But, of course, others have beat me to it, and I read in their track the hop, skip, sit a while, hop, skip of a rabbit, the unhurried gait of deer, and here where disturbed snow and a fluff of fur remember death by owl or hawk. Mine is the only noise in a world made silent by the falling snow, scrinching and crunching through a powder whose fluff absorbs all other sound. When I stop and listen, I hear nothing, nothing at all.

The old railroad bed along the Maury River lies flat and broad and begs for walking on, its turns inviting me to go round just one more bend to see what's there. What's there's another bend, and another bend beyond that. And the snow, falling silent in the fields and forest, lovely, dark, and deep, piling precarious upon branch and fence wire, waiting to shake down my neck. Like everyone possessed by snowy woods, I walk too far. Tired from the snow that clings to my boots as if to ask me to stay a little longer, from the cold that seeps through glove and scarf and sock, I think me in Jack London's woods and cast a wary eye on snow-laden hemlock branches under which I'd never build a fire. I spit; there's no crackle here in air that already wants to melt the snow. Yet I catch a flash of red through the world of white: a cardinal that thinks itself Sergeant Preston of the RCMP flies off to rescue snow blinded snowbirds.

Save for a now-gone bird, the world is black and white, a photograph before Kodak colorized the universe, back when, surveys suggest, more people dreamed in black and white than color. Now we see Technicolor even in our dreams and think ourselves possessed of knowledge bordering on wisdom. But I am cold, cold and tired and a bit farther from the car than I had thought I'd be. Evening is coming on, and I turn back, my tracks already disappearing beneath the snow that falls, falls, falls, falls, falls. I think the world white when I spy, through the screen of naked sycamore, snow vanishing into the river, slate gray and gelid, and sense, in my bones, all that I know, all that I am, is snow falling on water.

# MY GRANDCHILDREN'S FOREST

Gray ghosts of gone worlds haunt the woods we walk. Even if we didn't know we tread the graves of chestnut, elm, oak, hemlock, and ash, we'd know these trees we see today rose on the ruin of primeval forests we shall never see. Nor our children, nor our children's children. What trunks will they in their childhoods climb, what shades haunt their older, woodland walks, what trees provide the coffins they sleep in? What do the entrails of a dozen fallen monarchs reveal to botanizing high priests?

The end of southern forests of spruce and fir. My son's son and daughter may never stand upon Mount Roger's spruce- and fir-cloaked summit and wonder what the world below looks like. In their day, they will see Virginia laid out through the dark glass of a graveyard of broken trunks, the ruined end of Fraser's fir, whose prophecy I saw atop Mount Mitchell years ago. If not the adelgid, then acid rain, or global warming. His sister and he may, like me, make sock sachets of scavenged balsam needles, but these will come from trees grown in rows like vegetables. I've seen the ghost of Christmas yet to come in the trimmed-perfect trees of farms, which may supply a tree but can't replace the traipse through field and forest with an axe in search of something God and wind and storm shaped with an eye less to geometry than mystery.

The shrinking of the cornucopia that is the Appalachian cove. They will endure, these triumphs to 200 million years of change, but even more reduced than today's ragged scraps of forests once too deep and dark to be believed. Development, which sounds a lot like devilment, will continue its mad pace devouring what we haven't yet destroyed. The Midwest will still prefer to let its acid rains fall elsewhere than round the towns whose manmade plants poison nature's greener, grander plants. And some say global warming's hot dry breath will so desiccate these mountains that, where I walked through fields of

knee-high trillium beneath the shade of green- and yellow-flowered tulip poplars and breathed the honeyed basswood's scent, lousewort will bar our children's children path beneath a forest not yet born of drought-defying oak and
hickory. Which may, by our great-great-grandchildren's day, be large enough
and old enough to help the world forget the miracle of spring ephemera.

You cannot kill them all, the scoffers cry. Perhaps they're right this time—
unlike their great grandparents who cried the same at those who would stop
the dying of the passenger pigeon, the Carolina parakeet, the ivory-billed
woodpecker. Perhaps not—and, if not, then our great great grandchildren
may, as I did in the museums of my childhood, stare at the extinct in their glass
cages and sense, if only dimly, how diminished is the world greed left them.

Ironically, the same greed that condemns to death our natives supplies
aliens that will grow in their place, many with roots older than we who call ourselves American. The Asian tree-of-heaven, *Ailanthus altissima*, about which is
nothing heavenly but its height, has spread like a weed since its introduction
here in 1784. Virginia's farmers curse its ill-scented volunteers that sprout
like dragon's teeth in field and fencerow. Hike miles into the wilderness, and
there, accompanied by dandelion and plantain, stands in a clearing an *Ailanthus* grove—so popular with Victorians taken by the plant's purported tropic
silhouette they coined the term "ailanthery." My town neighbors felled one in
their backyard and offered the sectioned trunk as free wood to passersby. And
had no takers, the wood so fast grown and light it's not worth burning unless
you're building bonfires for the damned.

You can still buy from catalogues the prettier and perfumed Princess tree,
Pawlonia, whose wood the Japanese pay handsomely for. Less invasive than
*Ailanthus, Pawlonia,* also from the Far East, has found herself a home in the
Appalachians. Biking the Parkway in spring can delight a nose when it comes
upon a big-leaved *Pawlonia* sprout from the rock and in full bloom, its pale
purple blossoms sweet as the honeysuckle wrapped around its trunk. Invasive thickets of *Lonicera* strangle what they can, though none's as impressive as the equally Oriental kudzu, whose government-encouraged use as a
natural erosion control has swaddled miles of mountain roadsides in a green
garrote.

And why this litany of Asiatic horrors? Because China, the Koreas, and
Japan enjoy climates similar to ours and so have evolved plants happy to
colonize our East Coast. Having left in Asia their co-evolved herbivores and

pathogens, they spread here as rapidly and disastrously as did our ancestors. American and European plant explorers brought back a garden's worth of plants we treasure—azaleas, buddleia, camellias, gardenias—as well as plants less liked—bamboo, kudzu, honeysuckle. Sometimes we invite in the pathogen and leave the host behind. Before the Ice Ages, a great temperate forest wrapped round the Northern Hemisphere, and many American species have close relatives in Asia. Drifting continents and grinding glaciers long severed family ties. But the plants are close enough that a pathogen that feeds on one might feast on a kindred plant. As did the blight that killed our elms and chestnut, as do the bugs now killing our ash and hemlock. Globalization promises to bring even more, its wooden pallets and crates the unintended sneakers-in of invasive insects, bacteria, fungi, and viruses. As unstoppable as the zebra mussels devastating the Great Lakes imported in ships' ballast. Or the American jellyfish that almost killed the Black Sea.

Two hundred years before our forests suffered from an invasion of the Asiatic, they faced the frenzy of European weeds. Six thousand years of proximity to agriculture taught dandelion, plantain, purslane, chickory, and Queen Anne's lace the secrets of survival. Loosed on a New World, they did as much damage as their two-footed collaborators. Take a picture of yourself standing on your lawn, and you have a Wanted Dead or Alive poster for the most invasive plant and animal in North America—garden grass and *Homo sapiens*. No kudzu-choked hillside, no river cursed with purple plantain, no spotted knotweed waste compares to the green deserts of our yards in terms of water waste, herbicide and fertilizer, gasoline and lawn mower. And no zebra mussel, nutria, mustang, or gypsy moth has, like *Homo sapiens*, driven to extinction a single species or threatened to erase entire ecosystems, such as our prairies and our longleaf stands. The despised chestnut blight decimated but a single species, after all. We humans have killed off no one knows how many.

Would you save the forests? You must then kill the very grandchildren you wish to save them for. A Faustian bargain I'll not sign. Theirs will be forests vastly different from those I walk. And while I would preserve as much of these as possible, I too walk woods my parents' parents would find diminished. My father told me of the now nearly-gone great longleaf groves of his youth, where he could ride a horse for miles through trees larger round than the pillars of a courthouse, of the swamps he boated filled with cypress older than Roman Empire. And he knew them well; as a boy he worked the railroad cut through the woods to harvest them. That railroad disappeared with the trees it carried, its bed now a trail through what the government declares a wilderness. The Valley's mountains are as paradoxical; Timber Ridge and James Face wildernesses were logged out at the same time as my father's forests were. The Reverend Horace Dowdy remembers his daddy logging these and has shown me the railroad's tracks still running through the forests. And many a trail I've hiked is a road the loggers cut to take the trees downhill. To this we in this country've come, that wilderness is now an eighty-year-old logged-out leftover. Yet, if I can stand upon a once-ravaged ridge and think myself alone with wilderness, deceived though I be, there's hope my grandson and daughter too will find, in whatever woods grow here in their day, the same solace I find in mine.

# About the Author

JOHN LELAND is the author of several books published by the University of South Carolina Press including *Aliens in the Backyard: Plant and Animal Imports into America, Learning the Valley: Excursions into the Shanandoah Valley,* and *Porcher's Creek: Lives between the Tides.* Leland teaches English at Washington & Lee University in Lexington, Virginia.